A DELTA TRANSFORMED

ecological functions, spatial metrics, and landscape change

IN THE SACRAMENTO-SAN JOAQUIN DELTA

A Report of the Delta Landscapes Project:
Management Tools for Landscape-Scale Restoration of Ecological Functions

PREPARED FOR THE CALIFORNIA DEPARTMENT OF FISH AND WILDLIFE
AND ECOSYSTEM RESTORATION PROGRAM

OCTOBER 2014

PROJECT DIRECTION
Robin Grossinger
Letitia Grenier
Julie Beagle

PRIMARY AUTHORS
April Robinson
Sam Safran
Julie Beagle

CONTRIBUTING AUTHORS
Robin Grossinger
Letitia Grenier

SPATIAL ANALYSIS
Sam Safran
Julie Beagle

DESIGN AND LAYOUT
Ruth Askevold
Sam Safran

SFEI
AOSOC

PREPARED BY San Francisco Estuary Institute-Aquatic Science Center

IN COOPERATION WITH AND FUNDED BY California Department of Fish and Wildlife
Ecosystem Restoration Program

SFEI-ASC
PUBLICATION #729

THE SACRAMENTO-SAN JOAQUIN DELTA
modern waterways, islands, and tracts

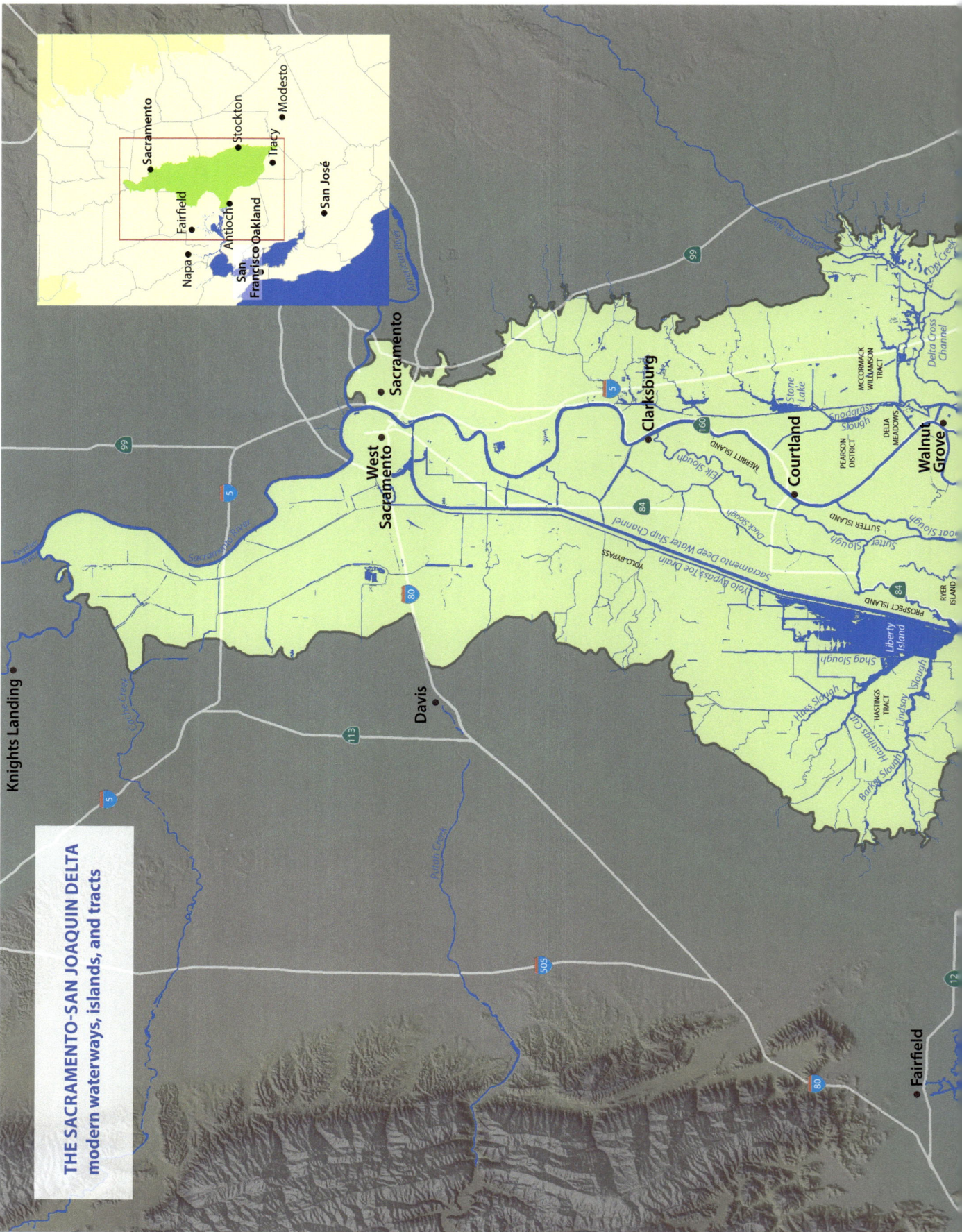

Sacramento

West Sacramento

Clarksburg

Courtland

Walnut Grove

Knights Landing

Davis

Fairfield

Sacramento River

Cache Creek

Putah Creek

YOLO BYPASS

Yolo Bypass Toe Drain

Sacramento Deep Water Ship Channel

Elk Slough

Sutter Slough

Steamboat Slough

MERRITT ISLAND

SUTTER ISLAND

PROSPECT ISLAND

RYER ISLAND

Liberty Island

Shag Slough

Hoss Slough

HASTINGS TRACT

Hastings Cut

Lindsay Slough

Barker Slough

Cache Slough

Miner Slough

Snodgrass Slough

Stone Lake

MCCORMACK WILLIAMSON TRACT

DELTA MEADOWS

PEARSON DISTRICT

Elk Slough

Steamboat Slough

Cosumnes River

Delta Cross Channel

Sacramento

Stockton

Modesto

Tracy

Fairfield

Antioch

Napa

San Francisco

Oakland

San José

American River

San Joaquin River

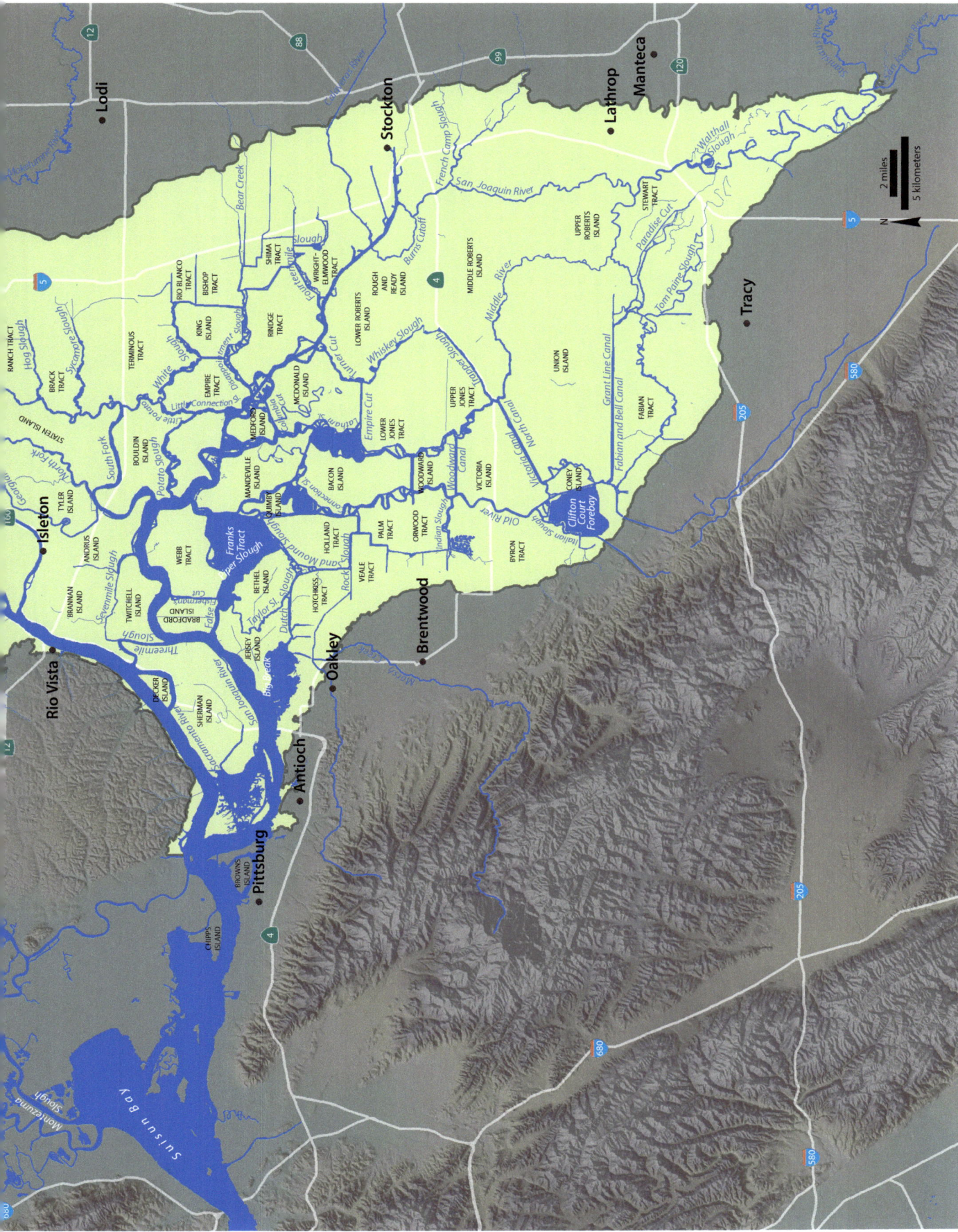

Lodi

Stockton

Lathrop

Manteca

Tracy

Isleton

Rio Vista

Antioch

Pittsburg

Oakley

Brentwood

2 miles
5 kilometers

N

RANCH TRACT
Hog Slough
Sycamore Slough
BRACK TRACT
STATEN ISLAND
North Fork
South Fork
Georgiana Slough
TYLER ISLAND
ANDRUS ISLAND
BRANNAN ISLAND
TWITCHELL ISLAND
SEVENMILE SLOUGH
DECKER ISLAND
SHERMAN ISLAND
BROWNS ISLAND
CHIPS ISLAND
Suisun Bay
Montezuma Slough

TERMINOUS TRACT
RIO BLANCO TRACT
BISHOP TRACT
KING ISLAND
White Slough
Little Potato Slough
Potato Slough
EMPIRE TRACT
BOULDIN ISLAND
WEBB TRACT
Franks Tract
Upper Jones Tract
BRADFORD ISLAND
BETHEL ISLAND
JERSEY ISLAND
Threemile Slough
False River
Fisherman's Cut
Taylor Sl.
Dutch Slough
Sandmound Slough
Big Break
Sacramento River
San Joaquin River

SHIMA TRACT
WRIGHT-ELMWOOD TRACT
Fourteenmile Slough
Disappointment Slough
RINDGE TRACT
MCDONALD ISLAND
Turner Cut
MEDFORD ISLAND
Columbia Cut
MANDEVILLE ISLAND
QUIMBY ISLAND
Connection Sl.
HOLLAND TRACT
HOTCHKISS TRACT
VEALE TRACT
PALM TRACT
ORWOOD TRACT
BACON ISLAND
Indian Slough
Rock Slough
WOODWARD ISLAND
Whiskey Slough
Empire Cut
LOWER JONES TRACT

Bear Creek
French Camp Slough
San Joaquin River
Burns Cutoff
ROUGH AND READY ISLAND
LOWER ROBERTS ISLAND
MIDDLE ROBERTS ISLAND
UPPER ROBERTS ISLAND
Middle River
Latham Slough
UPPER JONES TRACT
Trapper Slough
VICTORIA ISLAND
Woodward Canal
Old River
North Canal
Victoria Canal
UNION ISLAND
BYRON TRACT
CONEY ISLAND
Clifton Court Forebay
Grant Line Canal
Fabian and Bell Canal
FABIAN TRACT
Italian Slough
Tom Paine Slough
Paradise Cut
STEWART TRACT
Walthall Slough

Mokelumne River
Old River

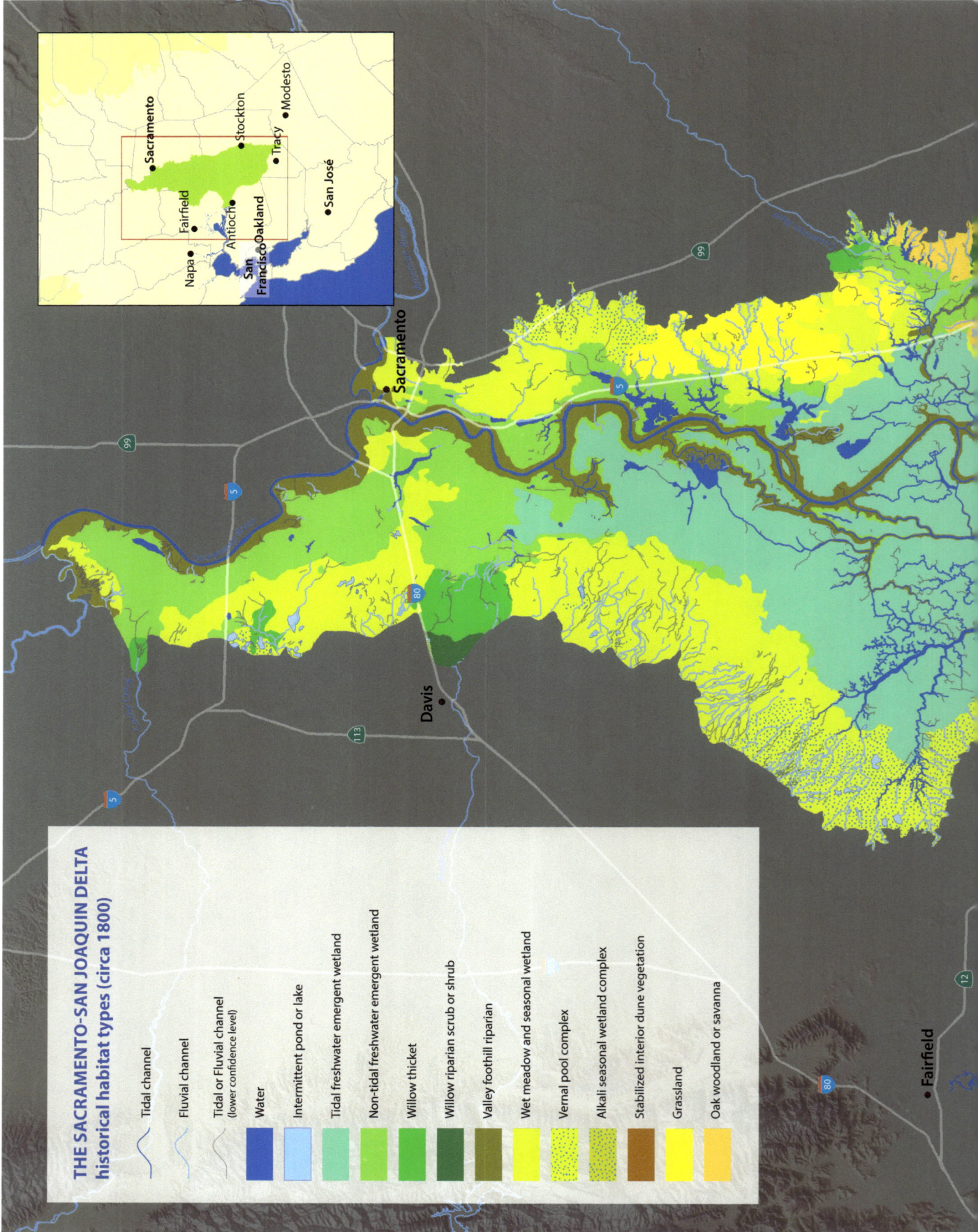

THE SACRAMENTO-SAN JOAQUIN DELTA
historical habitat types (circa 1800)

〰 Tidal channel

〰 Fluvial channel

〰 Tidal or Fluvial channel (lower confidence level)

▮ Water

▮ Intermittent pond or lake

▮ Tidal freshwater emergent wetland

▮ Non-tidal freshwater emergent wetland

▮ Willow thicket

▮ Willow riparian scrub or shrub

▮ Valley foothill riparian

▮ Wet meadow and seasonal wetland

▮ Vernal pool complex

▮ Alkali seasonal wetland complex

▮ Stabilized interior dune vegetation

▮ Grassland

▮ Oak woodland or savanna

Sacramento

Stockton

Modesto

Tracy

San José

Fairfield

Antioch

Oakland

San Francisco

Napa

Davis

Fairfield

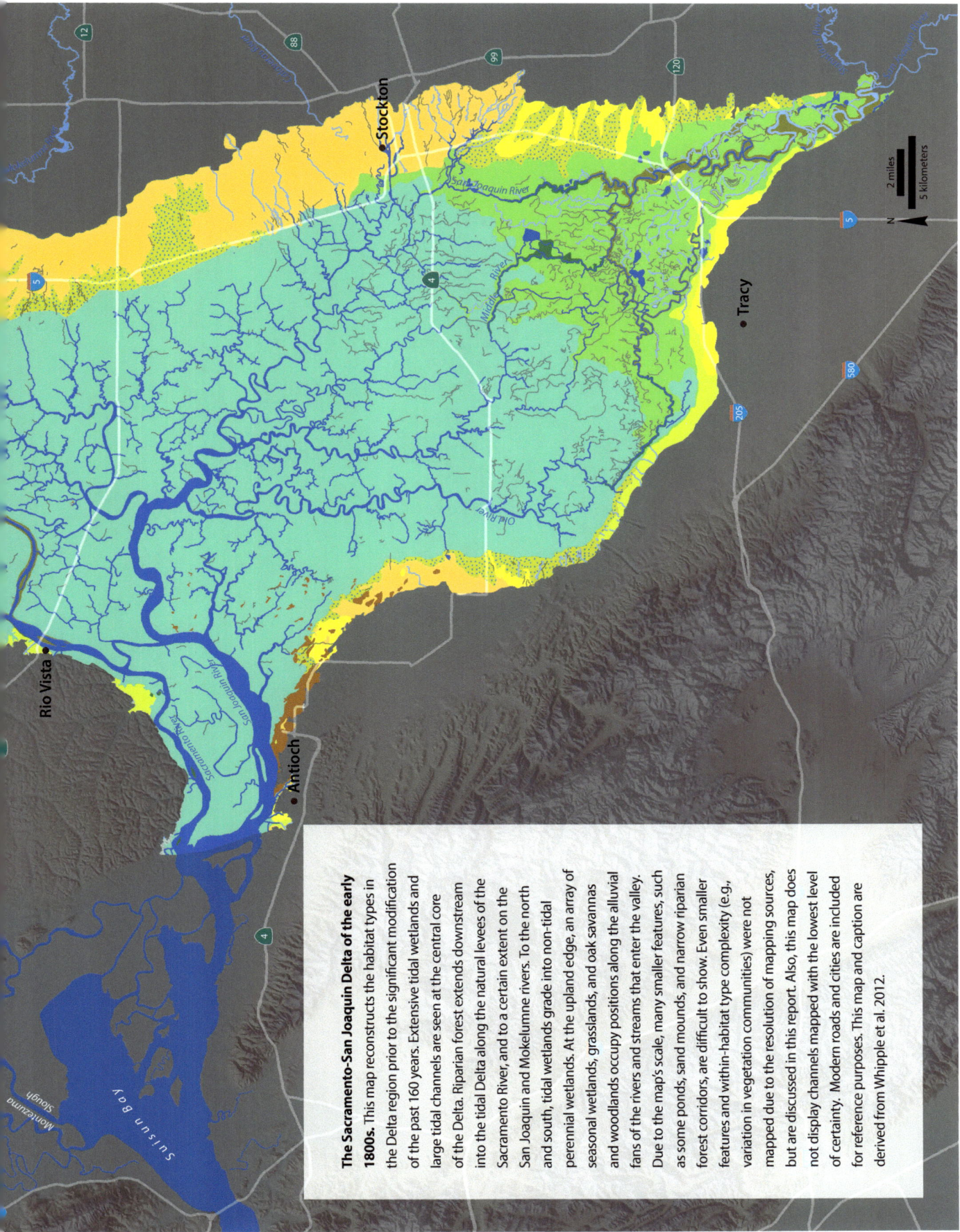

The Sacramento-San Joaquin Delta of the early 1800s. This map reconstructs the habitat types in the Delta region prior to the significant modification of the past 160 years. Extensive tidal wetlands and large tidal channels are seen at the central core of the Delta. Riparian forest extends downstream into the tidal Delta along the natural levees of the Sacramento River, and to a certain extent on the San Joaquin and Mokelumne rivers. To the north and south, tidal wetlands grade into non-tidal perennial wetlands. At the upland edge, an array of seasonal wetlands, grasslands, and oak savannas and woodlands occupy positions along the alluvial fans of the rivers and streams that enter the valley. Due to the map's scale, many smaller features, such as some ponds, sand mounds, and narrow riparian forest corridors, are difficult to show. Even smaller features and within-habitat type complexity (e.g., variation in vegetation communities) were not mapped due to the resolution of mapping sources, but are discussed in this report. Also, this map does not display channels mapped with the lowest level of certainty. Modern roads and cities are included for reference purposes. This map and caption are derived from Whipple et al. 2012.

Stockton

Tracy

Rio Vista

Antioch

San Joaquin River

Middle River

Old River

Sacramento River

San Joaquin River

Montezuma Slough

Suisun Bay

2 miles

5 kilometers

N

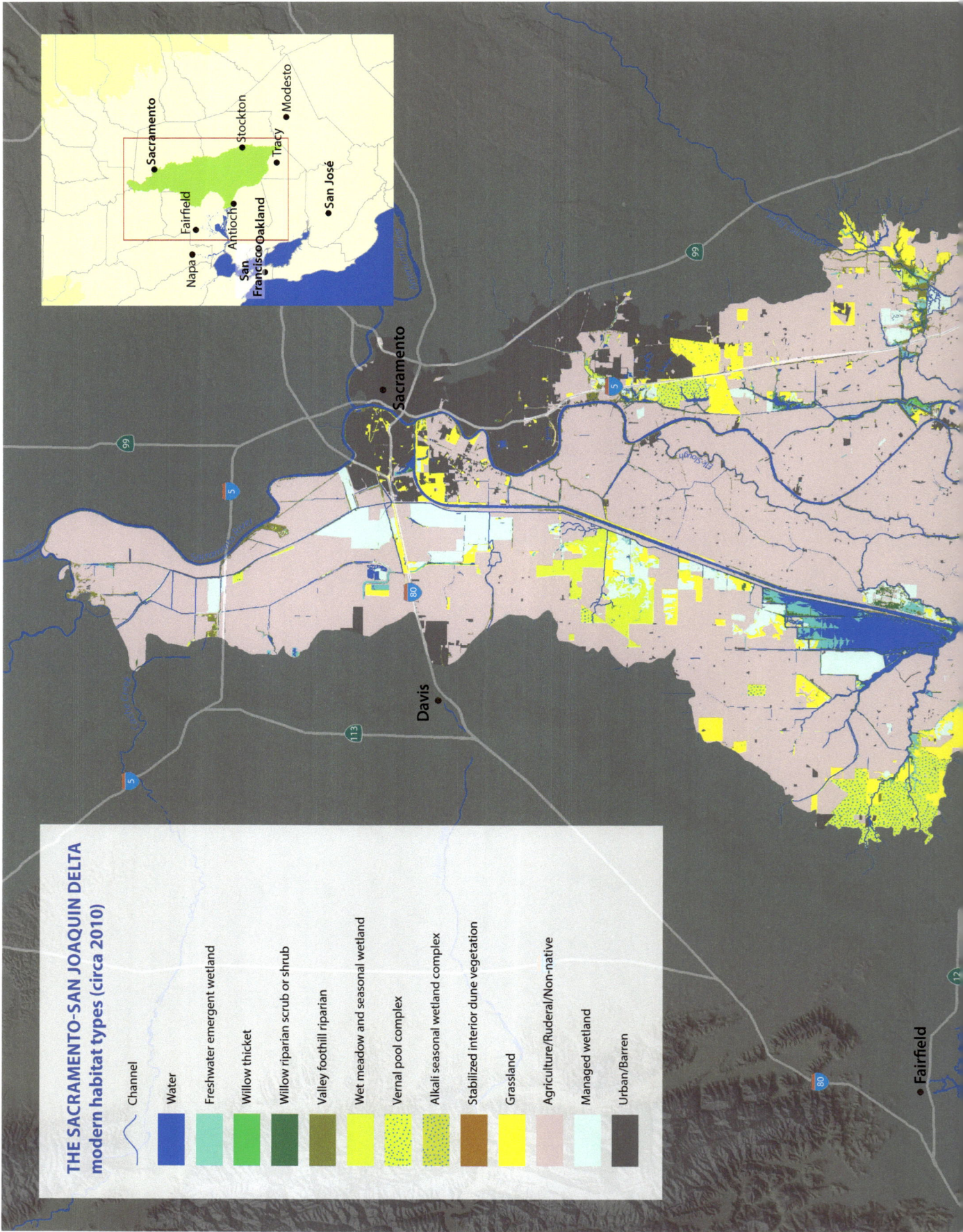

THE SACRAMENTO-SAN JOAQUIN DELTA
modern habitat types (circa 2010)

- Channel
- Water
- Freshwater emergent wetland
- Willow thicket
- Willow riparian scrub or shrub
- Valley foothill riparian
- Wet meadow and seasonal wetland
- Vernal pool complex
- Alkali seasonal wetland complex
- Stabilized interior dune vegetation
- Grassland
- Agriculture/Ruderal/Non-native
- Managed wetland
- Urban/Barren

Sacramento

Davis

Fairfield

Napa

Fairfield

San Francisco

Oakland

Antioch

Stockton

Tracy

Modesto

San José

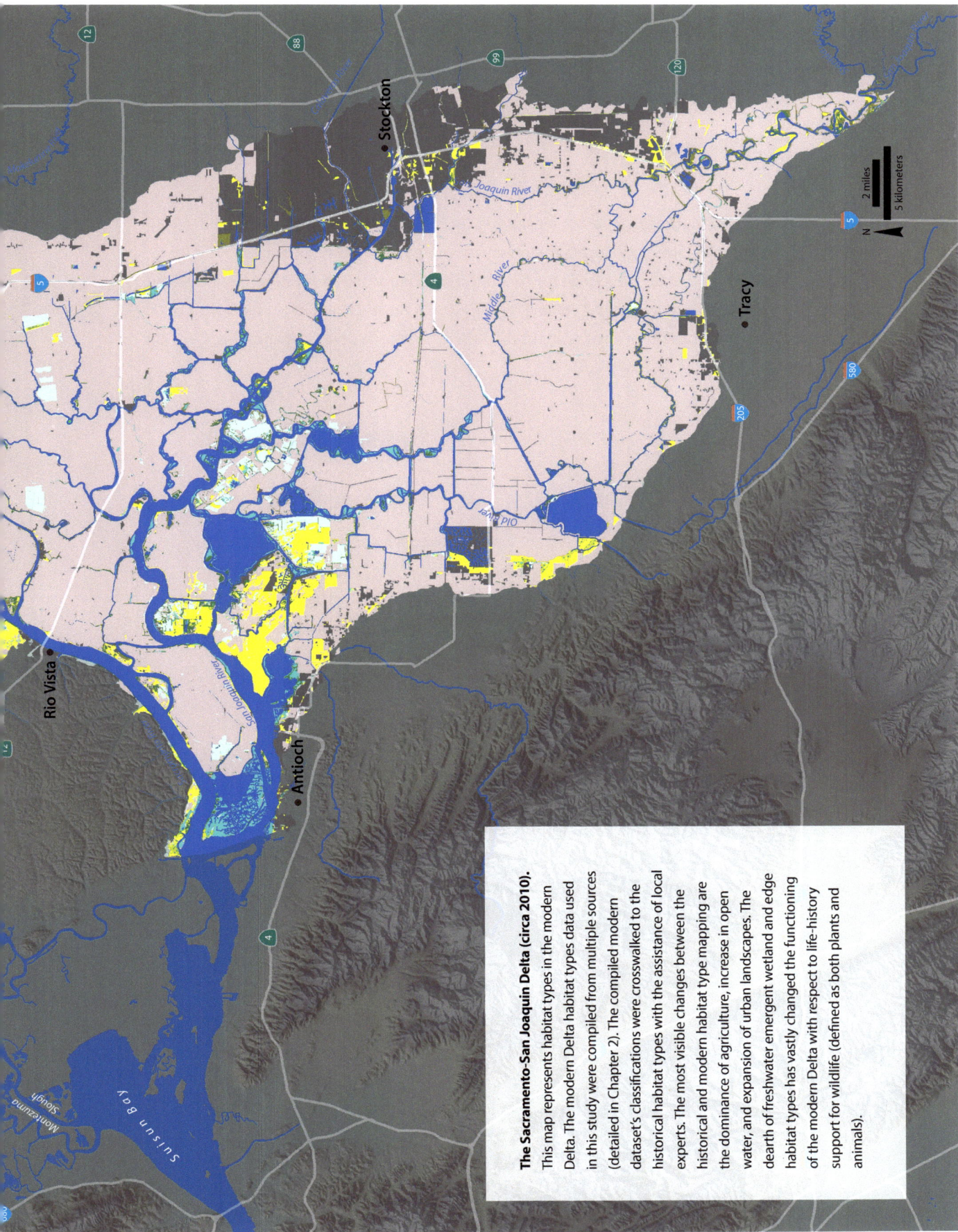

The Sacramento-San Joaquin Delta (circa 2010).
This map represents habitat types in the modern Delta. The modern Delta habitat types data used in this study were compiled from multiple sources (detailed in Chapter 2). The compiled modern dataset's classifications were crosswalked to the historical habitat types with the assistance of local experts. The most visible changes between the historical and modern habitat type mapping are the dominance of agriculture, increase in open water, and expansion of urban landscapes. The dearth of freshwater emergent wetland and edge habitat types has vastly changed the functioning of the modern Delta with respect to life-history support for wildlife (defined as both plants and animals).

SUGGESTED CITATION

San Francisco Estuary Institute-Aquatic Science Center (SFEI-ASC). 2014. A Delta Transformed: Ecological Functions, Spatial Metrics, and Landscape Change in the Sacramento-San Joaquin Delta. Prepared for the California Department of Fish and Wildlife and Ecosystem Restoration Program. A Report of SFEI-ASC's Resilient Landscapes Program, Publication #729, San Francisco Estuary Institute-Aquatic Science Center, Richmond, CA.

REPORT AVAILABILITY

Report is available on SFEI's website at www.sfei.org/projects/delta-landscapes-project.

COVER CREDITS

Front cover, left to right, top to bottom: maps of historical and modern marsh in the Sacramento-San Joaquin Delta (developed for this report); Liberty Island (photo by Barbara Beggs, USFWS).

Back cover, left to right: NAIP 2005; detail from map of historical habitats of the Sacramento-San Joaquin Delta (Whipple et al. 2012); detail from map of historical inundation in the Sacramento-San Joaquin Delta (Whipple et al. 2012); Snow and Ross' geese (photo by Steve Emmons, USFWS); portion of map by WH Hall, ca. 1880, Grand Island and Suisun Bay to foothills and 1st Standard North (Hall ca. 1880, courtesy of California State Archives).

CONTENTS

Acknowledgments

This project was funded by the California Department of Fish and Wildlife (CDFW) through the Ecosystem Restoration Program (ERP). We give special thanks to Carl Wilcox of CDFW and Cliff Dahm (former Delta Science Program [DSP] lead scientist), both of whom helped shape the project. We would like to thank Daniel Burmester, our CDFW project manager, for his invaluable technical advice and direction.

The project has benefited substantially from the sound technical guidance, engagement, and enthusiasm contributed by our Landscape Interpretation Team: Stephanie Carlson (University of California, Berkeley [UCB]), Jim Cloern (USGS), Chris Enright (DSP), Joe Fleskes (USGS), Geoff Geupel (Point Blue), Todd Keeler-Wolf (CDFW), William Lidicker (UCB), Steve Lindley (NOAA National Marine Fisheries Service), Jeff Mount (University of California, Davis [UCD]), Peter Moyle (UCD), Eric Sanderson (Wildlife Conservation Society), Anke Mueller-Solger (USGS), Hildie Spautz (CDFW), and Dave Zezulak (CDFW). Other key advisors to whom we are indebted include Brian Atwater (University of Washington), Jay Lund (UCD), and John Wiens (University of Arizona).

Alison Whipple (SFEI, UCD), Bill Fleenor (UCD), Stuart Siegel (Wetlands and Water Resources/ESA), Curt Schmutte (Metropolitan Water District [MWD]), Leo Winternitz (formerly of The Nature Conservancy), Peter Goodwin (DSP), Lauren Hastings (DSP), and Bruce Orr and Noah Hume (Stillwater Sciences) provided valuable insight and technical assistance along the way.

We give thanks to our separate team of advisors dedicated to discussing metrics relating to support for fish populations including: Carson Jeffres (UCD), Ted Sommer (California Department of Water Resources [DWR]), John Durand (UCD), and Jim Hobbs (UCD), as well as a group dedicated to waterbird life-history support: Dave Shuford (Point Blue) and Dan Skalos (CDFW).

Finally, we are grateful to many SFEI staff members (past and present) who contributed to this project: Sean Baumgarten, Erin Beller, Kristen Cayce, Jamie Kass, Marcus Klatt, Marshall Kunze, and Micha Salomon.

Executive Summary

While the decline of the Sacramento-San Joaquin Delta ecosystem is well recognized, relatively little is known about how the physical transformation of the Delta landscape—which took place more than a century ago—has affected its ability to support native plants and animals. The need for this understanding is urgent, as plans are being developed for substantial ecological restoration in the Delta. To fill this gap, we synthesized scientific knowledge about the Delta's native species with recent mapping of the pre-development (circa 1800) and contemporary Delta habitats to define ecologically relevant spatial metrics. We then analyzed the historical transformation of the Delta landscape from the perspective of these measures.

Based on scientific literature and input from experts, we identified aspects of the pre-development Delta landscape that contributed to the abundance and resilience of native wildlife populations. Habitats that dominated the landscape, such floodplains, marshes, and wide riparian forests, have declined precipitously in extent. For example, 98% of the freshwater emergent marsh in the Delta has been lost (from approximately 190,000 hectares to just over 4,000 hectares). Aquatic habitats have also undergone wholesale conversion. Underlying this habitat loss and degradation is the loss of the physical processes that create and maintain these habitats, conferring resilience upon the landscape, biological processes, and wildlife populations. The disconnection of floodwaters from marshes and riparian areas has not only altered habitats but also the exchange of materials and energy that affects the food web, water quality, and the future potential of these areas to be restored and provide habitat value. Thus, despite retaining some of the original system's template, with its sinuous channels and tidal flows, the Delta has been fundamentally transformed.

To improve the health and resilience of native wildlife populations in the Delta, another transformation will be required—one that restores greater habitat extent, connectivity, and diversity, as well as the physical processes that increase resilience and drive ecosystem function. This restoration must occur in the context of invasive species and changes in freshwater flow, necessitating a vision of the future that incorporates knowledge of the past and present but is completely new. This will require a landscape-scale framework for restoration that joins individual project "pieces" into a functional landscape "puzzle." The metrics presented here, as well as the landscape restoration conceptual models to be produced in the next phase of this project, can be useful tools to meet this challenge.

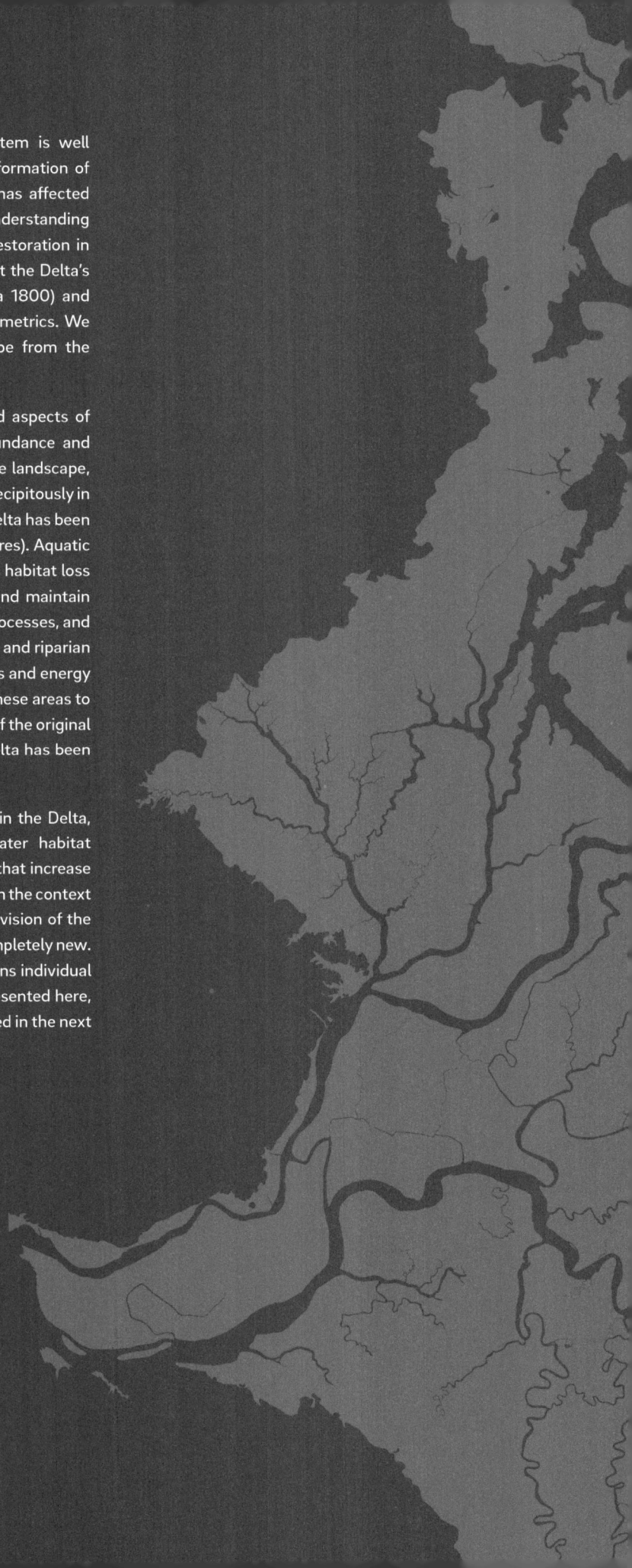

Recent state policy sets ambitious goals for ecosystem restoration in the Sacramento-San Joaquin Delta. The Delta Plan and California Water Code, as well as other regional documents, identify the need to go beyond small-scale habitat restoration to create larger functional landscapes of interconnected habitats.[1-6] Yet there is little quantitative guidance available to help design the complex spatial systems that are likely to achieve these goals. This report provides the first analysis of landscape ecology metrics in the pre-disturbance and contemporary Delta to help define, design, and evaluate functional, resilient landscapes for the future.

[1] California Water Code, Section 85302 (e)(1). "The Delta Plan shall include measures that...restore large areas of interconnected habitats within the Delta and its watershed by 2100."

[2] Teal et al. 2009. "Restoration strategies must be designed from a systems perspective that the Delta is considered as an interconnected watershed-river-marsh-estuary-ocean landscape."

[3] The Delta Plan 2013. "Management plans and decisions need to be informed by a landscape perspective that recognizes interrelationships among patterns of land and water use, patch size, location and connectivity, and species success."

[4] California Department of Water Resources 2013, Bay Delta Conservation Plan (BDCP; Public Draft). "The BDCP will contribute to the restoration of Sacramento-San Joaquin River Delta (Delta) ecosystems largely by addressing ecological functions and processes on a broad landscape scale."

[5] Wiens et al. 2012. "Historical ecology can provide a tool for using the past to understand the foundations of the present landscape and to assess its future potential for restoration by considering landscape patterns, processes, and functions and the conditions to which species are adapted."

[6] Delta Independent Science Board 2013. "We suggest that successful restoration projects in the Delta will [recognize that]... spatial context is part of the design. Individual restoration projects, regardless of their size, are not isolated from the surrounding aquatic and terrestrial landscape, or from restoration or management actions undertaken elsewhere."

1. Introduction

Delta Landscapes approach

Before modern development, almost half of California's coastal wetlands were found in the Sacramento-San Joaquin Delta. The Delta supported the state's most important salmon runs, the Pacific Flyway, and endemic species ranging from the delta smelt to the Delta tule pea. In the region's Mediterranean climate, the Delta's year round freshwater marshes were an oasis of productivity during the long dry season. Until reclamation, the Delta stored vast amounts of carbon in its peat soils. Today the Delta functions very differently, having undergone a massive and continuing transformation. Despite the dramatic changes, however, many native species are still found in the Delta, albeit in greatly reduced numbers. Some are threatened by extinction, and others may be soon.[1] The Delta no longer functions as a delta, spreading river and bay water and sediment across wetlands, floodplains, and riparian forests. Recovery of some of these lost ecological functions is considered crucial to ecosystem restoration in the Delta.[2]

Because of biological declines and regulatory challenges, Delta planning efforts often emphasize a few target species in habitat restoration and management. The Delta Landscapes project attempts to provide a "big picture" ecosystem perspective on how we reestablish ecosystem functionality for multiple suites of taxa. Our approach is to evaluate the landscape patterns and processes that supported native species in the historical Delta, measure how they have changed, and assess the potential for reestablishing smaller, modified, but ecologically functional deltaic landscapes in the future. The project contributes a missing dimension to Delta planning by providing a landscape-scale perspective on restoration opportunities that is founded in a sound understanding of how the Delta historically supported native species. This approach gives us the best chance at creating the new, reconciled landscapes of the future that integrate natural and cultural processes, maximizing resilience to climate change, invasive species, and other challenges.[3]

In order to imagine and plan for a functioning Delta ecosystem in the future, we must first understand how a healthy ecosystem looks.[4] Currently, we have no first-hand knowledge of how Delta landscapes functioned because there are no large areas typical of the historical Delta left. Such understanding is essential to evaluating the settings in which native wildlife (defined as plants and animals) evolved and designing future habitats that preferentially benefit these species. To develop this perspective, we analyzed early 1800s habitat mapping and other information from the Delta Historical Ecology Investigation,[5] completed in 2012, through a lens of key ecological functions that supported Delta wildlife. With a team of local and national experts in ecological and physical processes, we developed quantifiable metrics that represent different suites of functions

provided by different Delta settings. In order to evaluate change over time, the selected landscape metrics were also applied to the current Delta.

This first output of the Delta Landscapes project identifies important landscape-scale ecological functions that supported native species, and analyzes how they have changed. In subsequent project reports, these landscape metrics will be integrated with analyses of physical changes and existing constraints to explore the potential for future operational landscape units (OLUs) that would strategically link multiple projects over time into functional landscapes.[6]

Given the multiple uses of the Delta, diverse ecosystem stressors, and future challenges such as sea level rise and flooding, the future Delta will be a novel ecosystem,[7] likely to look very different from either the historical or the contemporary system. Today's Delta experiences multiple layers of impact, including freshwater flow diversions and alterations, contaminants, reduction in sediment supply, and non-native invasive species.[8] But while habitat mosaics cannot necessarily be reestablished in the same places or at the same scale at which they existed historically, they need to be designed to provide many of the same target functions at suitable scales. The challenge is to recognize of the potential resilience of disturbed physical and ecological systems, working in concert with underlying topographic and hydrological attributes to recover desired ecological functions.[9] By understanding how the landscape works and has changed, we can recognize the opportunities to strategically reconnect landscape components in ways that support ecosystem resilience to both present and future stressors.

Report structure

Following this Introduction, Chapter 2 presents a brief overview of the project framework and methods used (a longer, more detailed methods discussion is found in Appendix A). Chapter 3 discusses overall physical change in the Delta as it relates to ecological function. The next five chapters (Chapters 4-8) analyze different dimensions of life-history support for wildlife (animal and plants) in the Delta, focusing on particular habitat-associated guilds: fish, marsh wildlife, waterbirds, riparian wildlife, and marsh-terrestrial transition zone wildlife. Finally, Chapter 9 summarizes key findings and frames next steps in the Delta Landscapes project. The landscape analyses are presented as two-page spreads describing the selected ecological function, the spatial metrics used to evaluate that function, and analysis of that component of the landscape, past and present. Each of these chapters begins with several pages of preparatory background on the chapter topic.

The Delta landscape (below). Left to right: Nurse Slough, Sandhill Cranes at Stone Lake National Wildlife Refuge, giant garter snake, Sacramento River.

Photo Credits: Steve Martarano, USFWS; Justine Belson, USFWS; Brian Hansen, USFWS; Steve Martarano, USFWS

A short primer on the historical Delta landscape (summarized from Whipple et al. 2012)

The Sacramento-San Joaquin Delta historically served multiple physical and ecological functions. It was a perennial freshwater source in a Mediterranean climate, collecting, draining, and mixing water from the interior of the state (40% of the state's freshwater flows) to the ocean (**see map on pages iv-v**). It likewise served as an extended fluvial-tidal interface, with tidal influence extending past Sacramento. Saltwater influence was historically limited to the brackish Suisun marshes, and diminished towards Sherman Island, though the boundary was variable depending on the water year. Unlike coastal plain river deltas, the Sacramento-San Joaquin Delta is an inverted estuary that narrows at its outlet before opening to the San Francisco Bay.[10] It functioned as a sediment sink, slowing and settling coarser materials eroded from the granitic Sierras, while passing sands and silts downstream to replenish the salt marshes and beaches downstream. It was also the lungs of the region, sequestering carbon and releasing oxygen. The Delta was a highly productive system that provided abundant and diverse food resources to support robust food webs, including

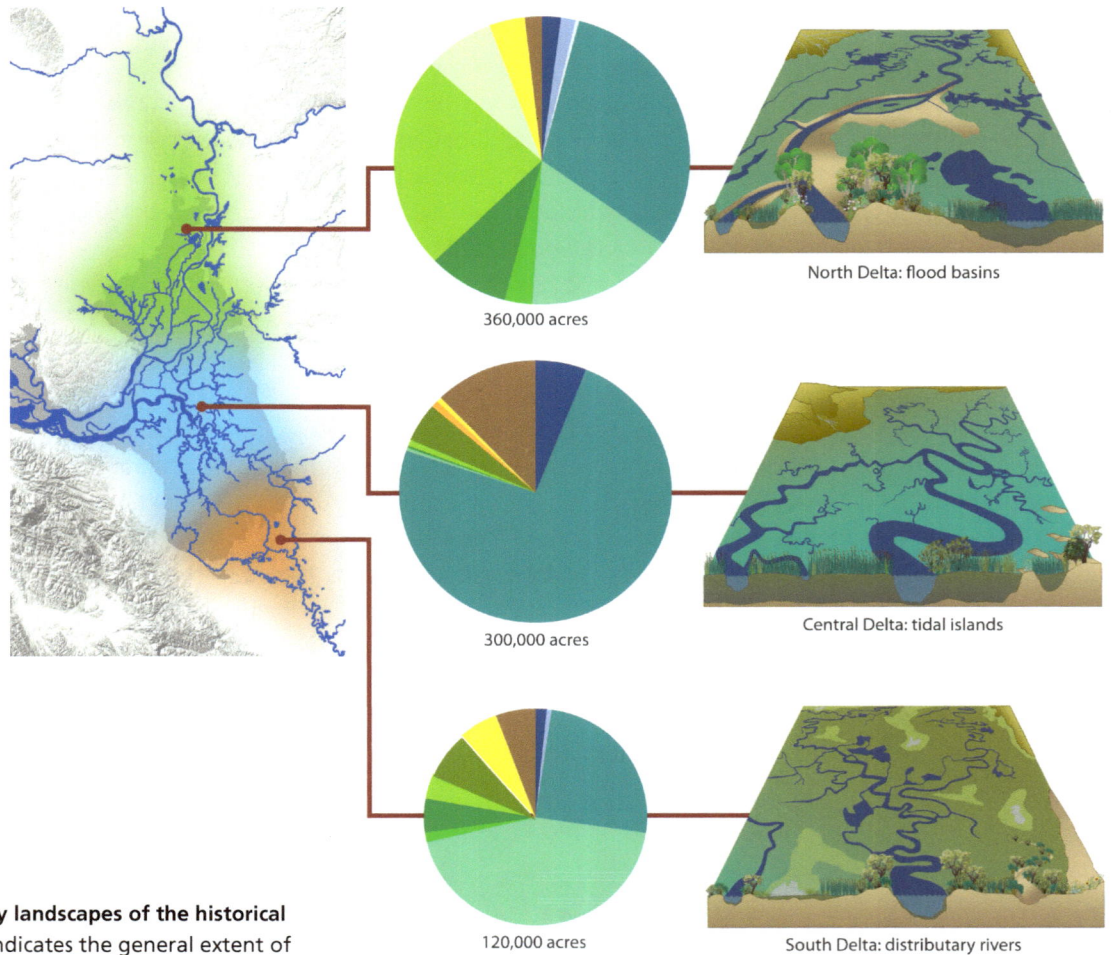

360,000 acres

North Delta: flood basins

300,000 acres

Central Delta: tidal islands

120,000 acres

South Delta: distributary rivers

The three primary landscapes of the historical Delta. The map indicates the general extent of the north Delta (a landscape of flood basins; shown in green), central Delta (a landscape of tidal islands; shown in blue), and South Delta (a landscape of distributary rivers; shown in brown). These landscapes were characterized by different assemblages and relative proportions of habitat types (as seen in the pie charts). Conceptual diagrams illustrating each of these landscapes are shown to right.

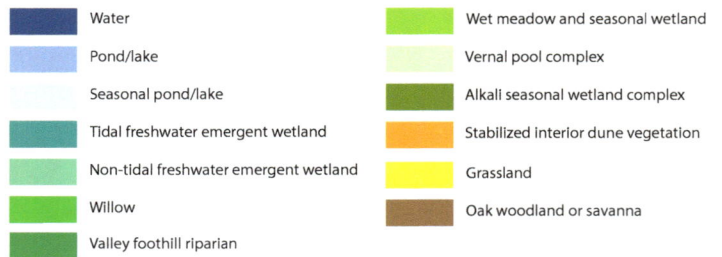

Legend:
- Water
- Pond/lake
- Seasonal pond/lake
- Tidal freshwater emergent wetland
- Non-tidal freshwater emergent wetland
- Willow
- Valley foothill riparian
- Wet meadow and seasonal wetland
- Vernal pool complex
- Alkali seasonal wetland complex
- Stabilized interior dune vegetation
- Grassland
- Oak woodland or savanna

indigenous tribes. Many native wildlife species were able to exploit the complex and resource-rich landscape of the Delta, some thriving in astonishing numbers.

The historical reconstruction of the Delta reveals the large-scale patterns and heterogeneity that existed before major anthropogenic influences.[11] The central, northern, and southern parts of the Delta were diverse in their geomorphic and hydrologic settings, and in the ecological functions they provided. The central Delta consisted predominately of islands of tidal freshwater emergent wetland (marsh), which supported a matrix of tule, willows, and other species. These wetlands—topographically almost flat—were wetted by twice daily tides, and inundated monthly (if not more frequently) by spring tides. During high river stages in the wet season, entire islands were often submerged with several feet of water. The large tidal sloughs had low banks and, like capillaries, bisected into numerous, progressively smaller branching tidal channels which wove through the wetlands, bringing the tides onto and off of the wetland plain, promoting an exchange of nutrients and organic materials. Channel density in the central Delta was greater than in the less tidally dominated northern and southern parts of the Delta (but lower than the brackish and saline marshes of the estuary downstream). The edges or transition zones around the central Delta were composed of alkali seasonal wetlands, grassland, oak savannas, and oak woodlands. On the western edge of the central Delta, sand mounds (remnant Pleistocene dunes) rose above the marsh, providing gently sloping dry land in an otherwise wet landscape that served as a high tide refuge for terrestrial species.

The ecological functions provided by the north Delta were driven primarily by the great Sacramento River, which created large natural levees and flood basins. These flood basins, running parallel to the river, accommodated large-magnitude floods, which occurred regularly, with inundation often persisting for several months. They consisted of broad zones of non-tidal marsh that had very few channels and transitioned to tidal wetland towards the central Delta. Dense stands of tules over three meters (m) (~10 ft) tall grew in these basins. Large lakes occupied the lowest points in these flood basins.

The north Delta's natural levees, created pre-Holocene by the large sediment supply of the Sacramento River, were broad, sloping features that graded into the marsh. These supra-tidal levees supported dense, diverse, multi-layered riparian forests often up to a mile in width. They ran parallel to the Sacramento River and other large tidal sloughs that conveyed enough sediment to build them over time during high flow events. The levees provided migration corridors for birds and mammals, and allochthonous input (organic debris) and shade to the river systems for aquatic species. Some areas within tidal elevations were seasonally isolated from the tides due to the presence of these levees and complex fluvial and tidal interactions. The edge of the north Delta was lined by seasonal wetlands and willow thickets, or "sinks," at the distal end of tributaries as they entered the flood basins.

The south Delta, like the north, was shaped by a large river system. Here, the three main distributary branches of the San Joaquin River created a complex network of smaller distributary channels, oxbow lakes, tidal sloughs, and natural levees of varying heights which graded across the long fluvial-tidal transition zone. In contrast with the single main channel of the Sacramento and the parallel flood basins, the San Joaquin River had less power and sediment supply to build high natural levees, and thus had many channels branching from the mainstem and coursing through the marsh islands; these channels vacillated between being fluvially or tidally dominated, depending on the time of the year. Small lakes and ponds were scattered in the south Delta, and the marsh was intersected with willow thickets, seasonal wetlands, and grasslands, making it a very diverse place for wildlife. The edge of the south Delta was dominated by alkali seasonal wetland complex, grassland, and oak woodland. The eastern edge of the Delta was shaped by the alluvial fans of the Mokelumne and Calaveras rivers that spread into the marsh.

The Delta was not a static place. Though the positions of large tidal channels, natural levees, and lakes were relatively stable, the Delta would have looked very different depending on the year and season. Areas of marsh that were flooded with several feet of water by late winter could be dry at the surface by late fall. The Delta was a place of significant spatial and temporal complexity at multiple scales.

2. Project Framework and Methods

This chapter provides a brief summary of the project framework and tools developed to assess ecological functions in the historical and modern Delta. A more detailed discussion of the underlying mechanics of these tools (metrics) can be found in Appendix A.

The Landscape Interpretation Team

The challenging task of exploring landscape-scale Delta ecological functions, identifying and quantifying landscape metrics, and eventually generating restoration tools and principles necessitates the collective best professional judgment of a team of experts. For this reason, an interdisciplinary group of high-level scientists was assembled as part of the initial project conception to provide regular input and guidance. This group is referred to as the "Landscape Interpretation Team" (LIT) and was drawn from relevant fields of expertise (including geology,

Landscape Interpretation Team members who have advised this project since its start.

LIT member	Affiliation
Stephanie Carlson	University of California, Berkeley
James Cloern	U.S. Geological Survey
Brian Collins	University of Washington
Chris Enright	Delta Science Program
Joseph Fleskes	U.S. Geological Survey
Geoffrey Geupel	Point Blue Conservation Science
Todd Keeler-Wolf	California Department of Fish and Wildlife
William Lidicker	University of California, Berkeley (Professor Emeritus)
Steve Lindley	National Oceanic and Atmospheric Administration/National Marine Fisheries Service
Jeff Mount	University of California, Davis
Peter Moyle	University of California, Davis
Eric Sanderson	Wildlife Conservation Society
Anke Mueller-Solger	U.S. Geological Survey
Hildie Spautz	California Department of Fish and Wildlife
Dave Zezulak	California Department of Fish and Wildlife

Other advisors: Brian Atwater (University of Washington), Daniel Burmester (CDFW), Jay Lund (UC Davis), John Wiens (Point Blue Conservation Science).

geomorphology, hydrodynamics, animal ecology, plant ecology, landscape ecology, and water resource management). Nineteen individuals have served on the LIT since the Delta Landscapes Project's initiation in 2012 (see table on previous page). LIT members have been consulted individually throughout the project and have met in plenary on five occasions. To date, the LIT has worked closely with SFEI-ASC staff to (1) identify ecological functions provided by the historical Delta's landscapes, (2) identify and prioritize landscape metrics that allow us to assess the extent and distribution of these key ecological functions (both historically and today) and (3) review/ interpret initial results.

Identifying key ecological functions provided by historical Delta landscapes

Functions summary

Using the guidelines described below, SFEI-ASC staff first developed a draft list of ecological functions likely provided by the historical Delta. Next, via an iterative process, the draft list was reviewed, prioritized, and edited by the LIT. The result—a final list of key ecological functions for the project to assess—is provided below and in the diagram on page 8. In this section, we also discuss our use of the term "ecological function," how we arrived at the ecological functions list, and each individual function.

POPULATION-LEVEL FUNCTIONS

Functions related to life-history support for wildlife

1) Provides habitat and connectivity for fish

2) Provides habitat and connectivity for marsh wildlife

3) Provides habitat and connectivity for waterbirds

4) Provides habitat and connectivity for riparian wildlife

5) Provides habitat and connectivity for marsh-terrestrial transition zone wildlife

Functions related to wildlife adaptation potential

6) Maintains adaptation potential within wildlife populations

COMMUNITY-LEVEL FUNCTIONS

Functions related to food webs

7) Maintains abundant food supplies and nutrient cycling to support robust food webs

Functions related to biodiversity

8) Maintains biodiversity by supporting diverse natural communities

---

LEVEL			**POPULATION**			**COMMUNITY**	
THEME		**Life-history support**			**Adaptation potential**	**Food webs**	**Biodiversity**
FUNCTION	CHAPTER **4** Provides habitat and connectivity for fish	CHAPTER **5** Provides habitat and connectivity for marsh wildlife	CHAPTER **6** Provides habitat and connectivity for waterbirds	CHAPTER **7** Provides habitat and connectivity for riparian wildlife	CHAPTER **8** Provides habitat and connectivity for marsh-terrestrial transition zone wildlife	Maintains adaptation potential within wildlife populations	Maintains food supplies and nutrient cycling to support robust food webs	Maintains biodiversity by supporting diverse natural communities
METRICS	Inundation extent, duration, timing, and frequency *p. 38*	Marsh area by patch size (patch size distribution) *p. 50*	Ponded area in summer by depth and duration *p. 61*	Riparian habitat area by patch size *p. 64*	Length of marsh-terrestrial transition zone by terrestrial habitat type *p. 72*	To be addressed with qualitative conceptual models in next phase of project.	Expected to be addressed with a related project.	To be addressed with qualitative conceptual models in in next phase of project.
	Marsh to open water ratio *p. 44*	Marsh area by nearest neighbor distance *p. 52*	Wetted area by type in winter *p. 61*	Riparian habitat length by width class *p. 66*				
	Adjacency of marsh to open water by length and marsh patch size *p. 44*	Marsh core area ratio *p. 54*						
	Ratio of looped to dendritic channels (by length and adjacent habitat type) *p. 46*	Marsh fragmentation index *p. 56*						

The ecological functions provided by the historical Delta and the metrics used to assess the extent and distribution of these functions. Functions were identified at both the wildlife population and community level, and were grouped into four themes. Although we describe each of these themes, only functions related to life-history support for wildlife are analyzed in detail for this report (see chapters 4-8). The other functions (shown with transparent colors in this diagram) will be assessed in future tasks and related projects.

What are ecological functions?

Much has been written on the meaning of the word "function" as it is used in the discipline of ecology.[1] In this report we use "functions" to mean "processes or manifestations of processes."[2] Smith et al. (1995) expand upon this basic definition and write that "wetland functions" are "the normal or characteristic activities or actions that occur in wetland ecosystems, or simply, the things that wetlands do. Wetland functions result directly from the characteristics of a wetland ecosystem and the surrounding landscape and their interaction."[3] By choosing to focus specifically on "ecological" functions, we adopt the general framework of the above definitions, but alter their focus. We define "ecological" as the relationship of organisms to one another and to their surroundings. For the purposes of this project, then, "ecological functions" are defined as the processes or manifestations of processes that support organisms. When we identify key ecological functions we are, in effect, attempting to answer the question: "how did the historical Delta environment support life?"

How did we choose which ecological functions to assess?

Environmental processes that support organisms occur at multiple scales, from global to microscopic, and almost any individual function can be broken down into component sub-functions.[4] The function *'provides suitable nesting habitat for Least Bell's Vireo,'* for example, is contingent on the function *'supports riparian vegetation communities with dense shrub cover,'* which, in turn is based on functions like *'promotes successful Salix* spp. *reproduction'* and *'maintains groundwater levels.'* If every process that supported Delta species were called out as a separate ecological function, the number of possible ecological functions would be effectively infinite. We were therefore required to identify and group ecological processes that supported Delta organisms into a manageable number of meaningful functions. To accomplish this, we established the following guidelines:

- **Focus on landscape-scale ecological functions.** We focused on capturing the degree to which specific ecological functions were provided by the overall landscape, and where in the landscape those functions were provided.

- **Focus on functions at both the population level and community level.** We desired to capture functions at both the population and community levels. For example, although food availability is a critical component of the ecological functions relating to population-level life-history support, we also sought to address Delta-wide productivity at the community level. Constraints on primary production and the relative importance of different production sources to the food web are major sources of uncertainty for Delta management today.

- **Focus on key ecological functions.** To keep this task manageable, we were required to focus on a limited number of key ecological functions—those that would have likely and collectively supported healthy wildlife communities in the Delta.

- **Focus on ecological functions for native wildlife.** Our focus on wildlife (which we define here as native plants and animals) is guided by the Delta's regional regulatory framework. The draft Bay Delta Conservation Plan (BDCP), for example, is designed in part to provide for the conservation and management of 56 covered plant and animal species. We focus much of our attention on vertebrates, since they tend to be better researched, are near the top of food webs, and are generally of greater interest to humans.

- **Consider life-history support functions for wildlife groups rather than for individual species.** For functions related to life-history support, we felt it necessary and useful to focus on specific ecological groupings. Ultimately, the ecological groupings we delineated for analyses were fish, marsh wildlife, waterbirds, riparian wildlife, and marsh-terrestrial transition zone wildlife. These groupings are largely based on habitat associations, which we felt was a sensible way to group species given the habitat-based GIS data we use for our analyses.

- **The extent and distribution of functions should be assessable through landscape metrics and supported by the available data.** We prioritized ecological functions for which appropriate landscape metrics and datasets were available to assess the function's extent and distribution (ideally both historically and today).

- **Focus on functions relevant to regional restoration efforts.** We prioritized ecological functions aimed to increase performance of the entire ecosystem, and used the framework of increased resilience and biodiversity to support the Delta's threatened and endangered species as specified by BDCP.

Function descriptions

Through a careful consideration of the historical habitat type map and discussions with the LIT, we identified eight key ecological functions of the historical Delta to focus on for this project (see the box on page 7 and the diagram on page 8). Functions can broadly be divided into four groups: those related to (1) wildlife life-history support, (2) wildlife adaptation potential, (3) food, or (4) biodiversity.

FUNCTIONS RELATED TO WILDLIFE LIFE-HISTORY SUPPORT The majority of this report focuses on wildlife life-history support functions. We define "life-history support" as the processes and characteristics of the Delta that supported the life histories of specific native taxa. Life-history support for wildlife encompasses many smaller species-specific functions, far more than could be detailed in this report. We therefore chose to focus on major wildlife groups: resident and migratory fish, marsh wildlife, waterbirds, riparian wildlife, and marsh-terrestrial transition zone wildlife. We assume that if the landscape provided broad life-history support for these groups then a majority of the related sub-functions were also being provided. Each of the functions related to wildlife support is described in the table below.

FUNCTIONS RELATED TO WILDLIFE ADAPTATION POTENTIAL For this report, "wildlife adaptation potential" is defined as the potential ability of native plant and animal populations to adapt to changing conditions. Wildlife adaptation potential encompasses adjusting to new or increased

The five functions related to wildlife life-history support. Each function relates to a specific wildlife group and is defined here with example sub-functions.

Function	Wildlife group	Description
Provides habitat and connectivity for fish	Native resident and migratory fish	Defined as the processes and the characteristics of the Delta that support the life histories of native resident and anadromous fish. Example sub-functions include *'provides sufficient floodplain inundation to support splittail spawning and rearing'* and *'provides adequate prey to support delta smelt.'*
Provides habitat and connectivity for marsh wildlife	Native marsh wildlife	Defined as the processes and the characteristics of the Delta that support the life histories of obligate and transitory marsh wildlife. Example sub-functions would include *'Black Rail refuge from predation'* (which would have been provided by dense vegetation) or *'tule seed germination'* (which would have been supported by inundation).
Provides habitat and connectivity for waterbirds	Native waterbirds	Defined as the processes and the characteristics of the Delta that support the life histories of waterbirds (which are defined as "birds that are ecologically dependent upon wetlands"[5]). Example sub-functions would include *'provides areas suitable for Sandhill Crane roosting,' 'provides food for wintering waterfowl,'* and *'provides nesting habitat for breeding ducks.'*
Provides habitat and connectivity for riparian wildlife	Native riparian wildlife	Defined as the processes and the characteristics of the Delta that support the life histories of riparian wildlife, including riparian residents and transients, particularly Neotropical songbirds. Example sub-functions would include *'provides nesting structures for riparian birds,' 'facilitates movement of terrestrial mammals,' 'provides food to avian fall migrants,' 'supports establishment of large valley oaks,'* and *'provides cover to anadromous fish in the form of large woody debris.'*
Provides habitat and connectivity for marsh-terrestrial transition zone wildlife	Native terrestrial-transition zone wildlife	Defined as the processes and the characteristics of the Delta that support the life histories of wildlife that utilize the transition zone between marshes and terrestrial habitats or these terrestrial habitats themselves. Example sub-functions would include *'provides tule elk with access to fresh water during the summer,' 'provides refuge to Black Rails during spring tides,' 'provides breeding pond habitat for California tiger salamanders.'*

disturbances and stressors, utilizing newly available resources, and moving as the locations of suitable conditions shift. Wildlife adaptation potential is particularly important in the face of climate change, sea-level rise, and changing water management in the Delta. Species distributions, habitat associations, and life-history strategies are likely to change over time in ways that are difficult to predict. Promoting wildlife adaptation potential at the landscape scale can help to manage for an uncertain future. The large population sizes with high genetic and phenotypic diversity that help drive adaptation potential require extensive, heterogeneous habitats. The ability of species to move along physical gradients (in elevation, salinity, and other parameters) as conditions change requires habitat connectivity. Metrics to characterize wildlife adaptation potential were not developed for this report, because this complex concept could not be adequately quantified with the resolution of data available. However, the drivers behind adaptation potential, namely habitat extent, connectivity, heterogeneity, and diversity, are integrated throughout this report (for example, the importance of alternative life-history support strategies for salmon is discussed in Chapter 4) and will inform future work on this project.

FUNCTIONS RELATED TO FOOD WEBS The amount of food within a system, and the ability of nutrients to be cycled and exchanged throughout that system, are critical to determining the degree to which that system can support wildlife. Constraints on primary production and the relative importance of different production sources to the food web are a major ecological uncertainty in the Delta system. We consider the size and location of high productivity habitats such as tidal marshes and shallow-water areas with high residency time to be important features for maintaining this function, and these are discussed in the related "life-history support" chapters. Estimating primary productivity in different parts of the Delta system was determined to be beyond the scope of this project, given the careful analysis of uncertainties that would be required. However metrics developed for this project may be appropriate to support such calculations in the future.

FUNCTIONS RELATED TO BIODIVERSITY For this project, we define biodiversity as the diversity of plants and animals supported by the Delta. Since biodiversity is the aggregate result of all the life-history support functions provided by the Delta, we do not devote a discrete chapter to biodiversity in this report. However, to understand changes in biodiversity at a landscape scale we make the following assumptions: 1) greater extent and diversity of habitat types will support greater diversity of species, 2) areas of key importance to endemic and rare native species are disproportionately important to overall biodiversity, and 3) preserving processes under which endemic species evolved may favor native over invasive species.

Identifying landscape metrics to assess ecological functions

What are landscape metrics?

Landscape metrics are commonly described as quantitative indices that describe spatial patterns of landscapes based on data from maps, remotely sensed images, and GIS layers.[6] McGarigal (2002) notes that "real landscapes contain complex spatial patterns in the distribution of resources that vary over time" and that "landscape metrics are focused on the characterization of the geometric and spatial patterns."[7] Landscape metrics are traditionally algorithms that quantify specific spatial characteristics of categorical data such as patches, classes of patches, or entire landscape mosaics. We broaden the term to use landscape metrics to quantify particular aspects of the physical landscape, including channel length, width and area, and habitat adjacencies in addition to analysis of patch dynamics. We use these landscape metrics to assess the extent and distribution of ecological functions. As such, the aspect of the landscape that the metric measures must somehow relate to the provision of the relevant ecological function.

Choosing landscape metrics

We used a series of rules to choose metrics that could be correlated to ecological functions and were feasible given the available data.

- **Landscape metrics are derived from the available data.** The selection of metrics was guided by the available data on the historical and present day Delta. The primary data sources for the historical Delta include a categorical map of historical habitat types and a linear network of historical channels and streams. Metrics were limited to those that could be derived from these and related contemporary data sources and were appropriate given the data's spatial extent and resolution.

- **Landscape metrics should be functional.** McGarigal (2002) uses the terms "functional" and "structural" to distinguish between metrics that measure landscape patterns with and without explicit reference to a particular ecological process.[8] Specifically, he defines functional metrics as "those that explicitly measure landscape pattern in a manner that is functionally relevant to the organism or process under consideration."[9] Since we are using landscape metrics to assess the extent and distribution of specific ecological functions, we selected only functional metrics. We conducted reviews of the available literature to parameterize our metrics for specific species/guilds of wildlife and to define how exactly the metrics relate to the functions they are meant to quantify. That said, some metrics intended to describe the physical landscape of the historical Delta are purely structural.

Metrics to assess the function 'Provides habitat and connectivity for fish'

1) Inundation extent, duration, timing, and frequency

2) Marsh to open water ratio

3) Adjacency of marsh to open water by length and marsh patch size

4) Ratio of looped to dendritic channels (by length and adjacent habitat type)

Metrics to assess the function 'Provides habitat and connectivity for marsh wildlife'

1) Marsh area by patch size (patch size distribution)

2) Marsh area by nearest neighbor distance

3) Marsh core area ratio

4) Marsh fragmentation index

Metrics to assess the function 'Provides habitat and connectivity for waterbirds'

1) Ponded area in summer by depth and duration

2) Wetted area by type in winter

Metrics to assess the function 'Provides habitat and connectivity for riparian wildlife'

1) Riparian habitat area by patch size

2) Riparian habitat length by width class

Metrics to assess the function 'Provides habitat and connectivity for marsh-terrestrial transition zone wildlife'

1) Length of marsh-terrestrial transition zone by terrestrial habitat type

---

Using the guidelines described above, SFEI-ASC staff first developed a draft list of landscape metrics that could be used to assess the extent and distribution of the ecological functions described above in both the historical and contemporary Delta. Next, via an iterative process, the draft list was reviewed, prioritized, and edited by the LIT and specialized expert groups. In addition to our meetings with the LIT, we also met separately with groups of regional experts to help review, vet, and parameterize the metrics chosen to assess specific functions. The result—a final list of landscape metrics for the project to analyze—is provided in the diagram on page 8 and in the box on page 12. For detailed descriptions of each metric and the methods used to execute them, please see Appendix A.

Calculating landscape metrics

Metrics were developed using spatial datasets of habitat types and channels/water bodies, both for the historical Delta (ca. 1800) and the modern Delta (ca. 2010). We used these layers to assess the chosen metrics for the entire Delta, both for the modern and historical periods. For more information on these datasets, please see the table and images on pages 14-15.

To best correlate our landscape metrics with ecological functions, we parameterized them based on relevant ecological thresholds and data identified in the available scientific literature (see table below). For certain metrics, categories or thresholds were identified to help make the results more easily interpretable in terms of ecological function. Examples of this include patch size, "large" patch size, and definition of "core" vs. "edge" habitat for marsh habitat. Although parameters are based on values from the literature, landscape metrics are inevitably simplifications of the complex relationships between habitat fragmentation and wildlife support, and do not necessarily account for important variables such as population demographics and habitat quality. Detailed information on sources and assumptions used to develop the metrics can be found in Appendix A.

Examples of sources and assumptions used to parameterize metrics (below). For each metric we present the parameter and the rationale used to justify it.

Metric	Parameter	Rationale
Marsh area by patch	When defining marsh patches, discrete marsh polygons were considered part of the same patch if they were located within 60 m of one another	This distance is derived from the rule set for defining intertidal resident rail (e.g. Black Rails) patches developed by Collins and Grossinger (2004), which is based on the best available data on rail habitat affinities and dispersal distances.[10] We assume that the rule set developed for intertidal rails in the South Bay (including Clapper Rails, which are not found in the Delta) is generally applicable to the Delta and non-tidal marsh. Additionally, this simplistic model of a binary landscape (marsh and non-marsh) assumes that all patches of marsh are equally suitable for rails, that the routes of travel between patches are linear, and that the only barrier to rail movement is distance.[11]
Marsh area by nearest "large" neighbor distance	Nearest "large" neighbor distance was calculated for each marsh patch as the linear distance to the nearest neighboring marsh patch of at least 100 ha.	This size threshold is based on (1) regression models of Spautz and Nur (2002) and Spautz et al. (2005), which show a significant negative correlation between Black Rail presence and distance to the nearest 100 ha marsh[12] and (2) the work of Liu et al. (2012), which found that Clapper Rail densities decrease in patches <100 ha.[13]
Marsh core area ratio	Core area ratio is defined as the percent of a marsh patch's total area that is greater than 50 m from the patch edge.	This distance is based on the work of Spautz and Nur (2002) and Spautz et al. (2005) indicating a significant positive relationship between Black Rail presence and marsh area >50 m from the marsh edge.[14]
Riparian habitat length by width class	We determined the length of riparian habitat in three width classes: <100 m wide, 100 – 500 m wide, and >500 m wide.	The 100 m width threshold is based in part on the work of Gaines (1974), who found that Western Yellow-billed Cuckoos were only present in patches at least 100 m wide.[15] Kilgo et al. (1998) found that riparian forest areas at least 500 m wide were necessary to maintain the "complete avian community" in bottomland hardwood forests in South Carolina.[16] These widths largely agree with the findings of Laymon and Halterman (1989) who (based on occupancy and nest predation rates) define riparian habitat <100 m wide as "unsuitable," habitats 100-600 m wide as "marginal" to "suitable," and habitats at least 600 m wide as "optimal" for cuckoo nesting.[17]

Datasets used to run landscape metrics. Data include habitat type layers, channel polygons, channel polylines, and channel bathymetry rasters. These layers were obtained or developed for both the historical and modern time periods.

Type of data	Time period	Notes
Habitat type (polygons)	Historical	The historical Delta habitat type data (A, right) used in this study were obtained from SFEI-ASC's *Sacramento-San Joaquin Delta Historical Ecology Investigation*.[18] The dataset classifies the historical Delta into 17 habitat types, the majority of which are based on modern classification systems. Some of these classifications were grouped to facilitate comparison with the modern Delta habitat types layer.
	Modern	The modern Delta habitat type data (B, right) used in this study were compiled from multiple sources, including the CDFW Vegetation Classification and Mapping Program's 2007 Sacramento-San Joaquin River Delta dataset[19] and the 2012 Central Valley Riparian Mapping Project Group Level dataset.[20] Together, these two sources covered greater than 99% of the project's study extent (C, right). The compiled modern dataset's classifications were crosswalked to the historical habitat types (or groups of historical habitat types) with the assistance of local experts.[21]
Channels (polygons & centerlines, bathymetry rasters)	Historical	Historical channel polygons (D, right) were obtained from SFEI-ASC's *Sacramento-San Joaquin Delta Historical Ecology Investigation* historical habitats layer by selecting polygons classified as 'fluvial low order channel,' 'fluvial mainstem channel,' 'tidal low order channel,' or 'tidal mainstem channel.'[22] Historical channel polylines were obtained from the *Delta Historical Ecology Investigation*'s historical creeks layer.[23]
		Historical bathymetry was derived from a variety of historical sources, including mid-19th century surveys of the Sacramento and San Joaquin rivers.[24] The task of developing a historical topographic-bathymetric digital elevation model of the Delta from these data is the focus of a separate project (a collaboration between the San Francisco Estuary Institute and researchers at the UCD Center for Watershed Sciences). This report utilizes interim data from that project.
	Modern	Modern channel polygons (E, right) were derived from the National Hydrography Dataset (NHD)[25] by clipping the dataset to the project study extent and selecting features classified as 'StreamRiver' or 'CanalDitch.' Additional channels that were not included in the NHD but are apparent in contemporary aerial photographs were either incorporated from other datasets (such as CDFW Delta LiDAR hydrography breaklines) or manually digitized by SFEI staff. Modern channel polylines were generated from the polygon dataset (described above) with a custom centerline generation tool.
		Modern bathymetry was extracted from a continuous topographic-bathymetric DEM of the San Francisco Bay-Delta Estuary.[26]

A HISTORICAL HABITAT TYPES

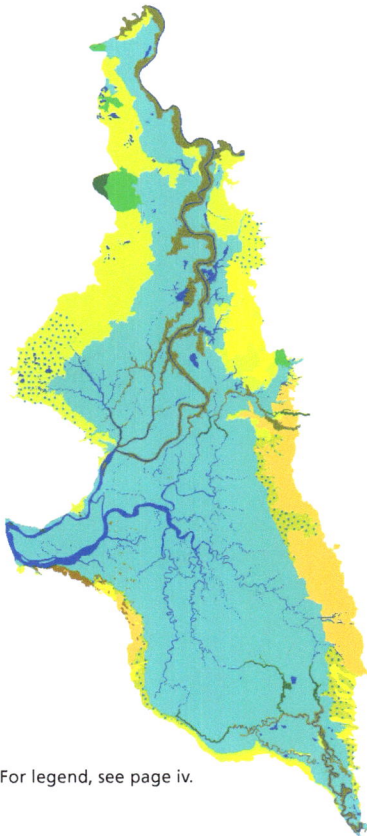

For legend, see page iv.

B MODERN HABITAT TYPES

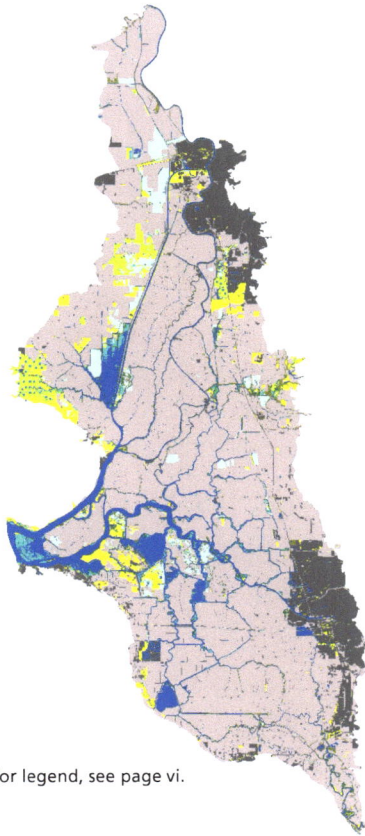

For legend, see page vi.

C MODERN DATA SOURCES

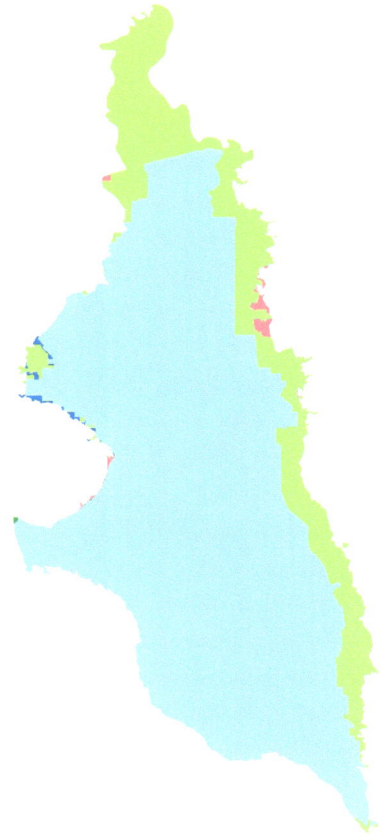

- CDFG 2007 – VegCAMP Delta Mapping
- CDWR 2012 – VegCAMP CVRMP
- SFEI 2013 – Supplemental Mapping
- WWR 2013 – CSCCA Natural Communities
- BDCP 2013 – Natural Communities

D HISTORICAL CHANNEL LAYERS

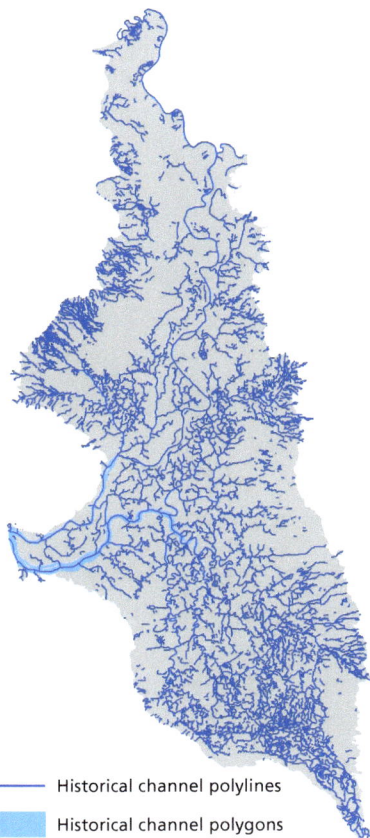

— Historical channel polylines

　Historical channel polygons

E MODERN CHANNEL LAYERS

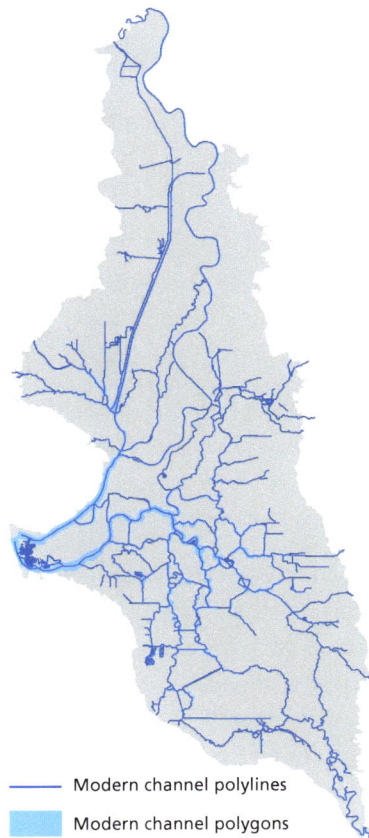

— Modern channel polylines

　Modern channel polygons

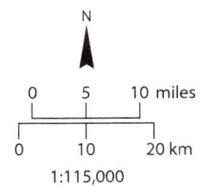

N

0　5　10 miles

0　10　20 km

1:115,000

Key project assumptions, limitations, and uncertainties

Inevitably, using available data sources for analyses of an ecosystem as complex as the Delta involves significant assumptions and uncertainties. Here we list the largest assumptions, uncertainties, and limitations associated with the use of our data. For more details, please refer to Appendix A.

General assumptions

Records of what wildlife were present in the historical Delta are sparse and inconsistent. Accounts of how wildlife used the landscape are even more so. Therefore, inferring the ecological functions provided by the historical landscape requires us to make many assumptions, with varying levels of confidence, combining disparate sources to develop a picture of the functioning landscape as a whole. Assumptions made and sources used are referenced in endnotes in the back of the report. Types of information, sources and assumptions used to interpret ecological functions in the historical landscape fell into several broadly defined categories:

- Assumptions based on well-established ecological theory.

- Assumptions based on ecological theory, but that required us to make major assumptions about Delta functioning. For these assumptions, the endnotes provide added detail on our rationale and sources.

- Assumptions based on ecological functions in less disturbed systems (e.g., salmon support in Pacific Northwest wetlands).

- Assumptions based on knowledge of natural history, physiological tolerance, and current habitat associations of Delta species.

- Assumptions based on records of historical occurrence. We did not go back to primary sources to look for incidents of species observations, but where these observations are summarized by other sources we cite them.

- Assumptions based on understanding of first principles of physical processes.

- Landscape metrics are a proxy for ecological function.

- Historical and modern habitat types are directly comparable.

Uncertainties (see Appendix A, pages 95-97 for additional details)

- **Uncertainty associated with the historical spatial data.** For the *Delta Historical Ecology Investigation,* each feature in the historical habitat types and channels layers was assessed for certainty. Overall, certainty of the features' interpretation/location was characterized as fairly high.[27]

- **Uncertainty associated with the modern spatial data.** Some degree of uncertainty is associated with each of the individual datasets compiled to generate our modern habitat types map. Additional uncertainty is associated with the process of crosswalking each of these data sources to the single classification system used in the historical dataset.

- **Uncertainty associated with historical and modern data fidelity.** When making comparisons between the historical and modern landscape, it was important that we compared the same things, at the same scale, using the same measurements. While, for certain analyses, differences in data resolution increased the uncertainty surrounding the precise magnitude of measured changes, we do not believe that these differences impacted the direction of changes or the overall stories told by the analyses.[28]

Limitations

- **The methods do not assess all of the functions that were performed by the historical Delta.**[29] Our high-level list of key ecological functions provided by the historical Delta is meant to broadly capture the functions that would have—likely and collectively—supported healthy wildlife communities in the Delta. Other high-level functions (such as primary productivity) are not addressed, while multitudes of lower-level functions (such as providing roosting habitat for certain bird species) are not specifically or directly identified in the body of this document. The project team decided which ecological functions to address using guidance from the LIT, who reviewed and edited a draft master list of possible ecological functions.

- **The metrics do not assess the landscape quantitatively for fine-scale heterogeneity.** Some historical and modern habitat types are mosaics that encompass smaller features (e.g., small ponds, beaver cuts, large woody debris, and willow-fern patches). We sometimes attempt to generally quantify these but do not discretely map or specifically analyze them.

- **The methods do not assess cultural, recreational, educational, or aesthetic functions of the historical (or contemporary) Delta.**[30] While there is limited information known about indigenous uses of the historical Delta, we recognize that humans had a significant impact on its ecological functioning. This is not a focus of this analysis.

- **Landscape metrics do not represent a direct measurement of the performance of a function.** Landscape metrics to represent ecological function are based on literature on conditions in California and elsewhere, but are not direct measurements of ecological function. As stated above as an assumption, metrics create a proxy for, or a hypothesis about expected ecological outcomes, based on observations elsewhere. The metrics do not include statistical validation/field testing.

- **Metrics do not capture interannual (or in some cases seasonal) variability in hydrology or temperature.** The data used for this analysis create a snapshot in time, from which we have inferred some seasonal and interannual variability. While seasonal variability is captured in timelines of available habitat through a water year for fish and waterbirds, the longer term interannual hydrologic patterns typical of our Mediterranean climate are not quantitatively assessed due to data limitations. Measurements of flow or sediment are not included.

- **The metrics do not acknowledge the limitations of private versus public land in terms of providing ecological function.** The analysis presented here does not distinguish between private or public land in the Delta. For restoration plans to eventually be made from these data, the details and constraints of land holdings must be considered.

- **The metrics do not differentiate between types of agriculture.** We recognize that certain types of wildlife-friendly agriculture are practiced in the Delta currently, and that certain crops and crop patterns provide more ecological benefit than others. At this scale of analysis, our report does not differentiate between types of agriculture, though further research could be done on this topic.

- **The report does not analyze the impact of invasive species or changes to groundwater levels on ecological functions.**

- **The metrics do not weight the modern land surface in terms of severity of subsidence.** During future stages of the Delta Landscapes project which involve integrating the results of the metrics into landscape units, these physical constraints will be considered.

Habitat type	Description
Water	**Tidal mainstem channel:** Rivers, major creeks, or major sloughs forming Delta islands where water is understood to have ebb and flow in the channel at times of low river flow. These delineate the islands of the Delta. **Fluvial mainstem channel:** Rivers or major creeks with no influence of tides. **Tidal low order channel:** Dendritic tidal channels (i.e., dead-end channels terminating within wetlands) where tides ebb and flow within the channel at times of low river flow. **Fluvial low order channel:** Distributaries, overflow channels, side channels, swales. No influence of tides. These occupy non-tidal flood-plain environments or upland alluvial fans. **Freshwater pond or lake:** Permanently flooded depressions, largely devoid of emergent Palustrine vegetation. These occupy the lowest-elevation positions within wetlands. **Freshwater intermittent pond or lake:** Seasonally or temporarily flooded depressions, largely devoid of emergent Palustrine vegetation. These are most frequently found in vernal pool complexes at the Delta margins and also in the non-tidal floodplain environments.
Freshwater emergent wetland	**Tidal freshwater emergent wetland:** Perennially wet, high water table, dominated by emergent vegetation. Woody vegetation (e.g., willows) may be a significant component for some areas, particularly the western-central Delta. Wetted or inundated by spring tides at low river stages (approximating high tide levels). **Non-tidal freshwater emergent wetland:** Temporarily to permanently flooded, permanently saturated, freshwater non-tidal wetlands dominated by emergent vegetation. In the Delta, occupy upstream floodplain positions above tidal influence.
Willow thicket	Perennially wet, dominated by woody vegetation (e.g., willows). Emergent vegetation may be a significant component. Generally located at the "sinks" of major creeks or rivers as they exit alluvial fans into the valley floor.
Willow riparian scrub or shrub	Riparian vegetation dominated by woody scrub or shrubs with few to no tall trees. This habitat type generally occupies long, relatively narrow corridors of lower natural levees along rivers and streams.
Valley foothill riparian	Mature riparian forest usually associated with a dense understory and mixed canopy, including sycamore, oaks, willows, and other trees. Historically occupied the supratidal natural levees of larger rivers that were occasionally flooded.
Wet meadow or seasonal wetland	Temporarily or seasonally flooded, herbaceous communities characterized by poorly-drained, clay-rich soils. These often comprise the upland edge of perennial wetlands.
Vernal pool complex	Area of seasonally flooded depressions, characterized by a relatively impermeable subsurface soil layer and distinctive vernal pool flora. These often comprise the upland edge of perennial wetlands.
Alkali seasonal wetland complex	Temporarily or seasonally flooded, herbaceous or scrub communities characterized by poorly-drained, clay-rich soils with a high residual salt content. These often comprise the upland edge of perennial wetlands.
Stabilized interior dune vegetation	Vegetation dominated by shrub species with some locations also supporting live oaks on the more stabilized dunes with more well-developed soil profiles.
Grassland	Low herbaceous communities occupying well-drained soils and composed of native forbs and annual and perennial grasses and usually devoid of trees. Few to no vernal pools present.
Oak woodland or savanna	Oak dominated communities with sparse to dense cover (10-65% cover) and an herbaceous understory.
Agriculture/Ruderal/Non-native	Cultivated lands, including croplands and orchards. This habitat type also includes areas dominated by non-native vegetation and ruderal lands.
Managed wetland	Areas that are intentionally flooded and managed during specific seasonal periods, often for recreational uses such as duck clubs.
Urban/Barren	Developed, built-up land often classified as urban, barren or developed. Includes rock riprap bordering channels.

Modern only

Photo by Daniel Burmester — Tidal mainstem channel

Photo by Bill Miller — Valley foothill riparian

Photo by Daniel Burmester — Fluvial mainstem channel

Photo by Daniel Burmester — Valley foothill riparian

Photo by Bill Miller — Fluvial low order channel

Photo by Daniel Burmester — Wet meadow/seasonal wetland

Photo by Bill Miller — Freshwater pond or lake

Photo by Marc Hoshovsky — Vernal pool complex

Photo by Daniel Burmester — Freshwater pond or lake

Photo by Jean Pawek — Stabilized interior dune vegetation

Photo by Bill Miller — Freshwater emergent wetland

Photo by Christopher Thayer — Stabilized interior dun3 vegetation

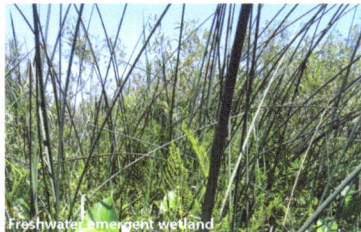

Photo by Daniel Burmester — Freshwater emergent wetland

Photo by Ingrid Taylar — Oak woodland or savanna

Photo by Bill Miller — Willow thicket

Photo by Jean Pawek — Grassland

Habitat types descriptions and images.
The mapping developed and used in this report includes twenty habitat types. With the exception of three types unique to the modern Delta, the classification was first developed for the *Sacramento-San Joaquin Delta Historical Ecology Investigation*.[31] The table **(opposite)** describes each habitat type. Representative images are shown to illustrate what these landscapes may have looked like. Not shown: alkali seasonal wetland complex, agriculture/non-native/ruderal, urban/barren, managed wetlands.

3. Overall Delta Landscape Changes

This chapter describes systemic changes to the Delta ecosystem since the historical period (prior to the analyses of ecological function in the subsequent chapters).

The historical Delta is gone. The defining characteristic of the historical Delta was its extensive wetland landscape, formed over time as floodwaters met the tides. Modern land management has increasingly disconnected floodwaters from the wetlands by widening and deepening channels, diking and draining wetlands for agriculture, and building levees for flood protection. The consequences of this disconnection include a nearly complete loss of Delta wetlands, along with the processes that sustain them, and a dramatic altering of the remaining aquatic habitats. The Delta has become more susceptible to invasive species, and the consequences of those invasions are magnified as a result of habitat loss and alteration. The ecological impacts of these transformations have been dire; the Delta food web has collapsed, wildlife populations have been drastically reduced in size, and the resilience of many remaining populations has been impaired.

The Delta once supported numerous wildlife species, some in great abundance, many of which are now species of concern. Tricolored blackbirds formed the largest breeding colonies of any landbird in North America,[1] Chinook salmon runs were among the largest on the Pacific Coast,[2] despite being at the southern end of the species distribution, and millions of waterfowl wintered in the Central Valley, in concentrations unmatched anywhere in California.[3] Many regionally endemic species inhabited the Delta, including plants (Mason's lilaeopsis, Delta tule pea), insects (Lange's metalmark butterfly, valley elderberry longhorn beetle), fish (delta smelt, longfin smelt, thicktail chub), reptiles and amphibians (giant garter snake, California tiger salamander), and mammals (riparian brush rabbit, riparian woodrat). At least one species endemic to the Delta, the thicktail chub, is now extinct, while several more have been extirpated in the Delta (including the Western Yellow-billed Cuckoo and Sacramento perch). Many more Delta species are at risk of being lost in the future; the draft Bay Delta Conservation Plan (BDCP) lists 56 species as being of immediate management concern.[4]

Six interrelated drivers of change are implicated in the loss of ecological function in the Delta. These drivers interact in a complex physical and biological system, where one driver may tip the scales toward ecosystem collapse, but only because the other drivers have brought the system to that tipping point.[6] The drivers of change are (1) reduction in habitat extent, (2) loss of heterogeneity within habitats, (3) loss of connectivity within and among habitat types, (4) degradation of habitat quality, (5) disconnection of habitats from the physical processes that form, sustain, and confer resilience upon them, and (6) invasion by ecosystem engineers such as Brazilian waterweed and invasive clams, and other predatory fish. Other drivers of change, particularly reductions and alterations in freshwater inflow and contaminants, are also responsible for the loss of ecological function.[5]

The habitats that dominated when the Delta was a functionally intact ecosystem have been reduced to small fractions of their former extent. For example, 15,608 hectares of Valley foothill riparian forest throughout the historical Delta have been reduced to 4,010 hectares: a reduction of 74%. There were at least 3,217 km of small channels (<15 m wide) in the Delta historically (not including an estimated 1,931 km of additional unmapped channels; see Appendix A, page 85), but only 144 km of small channels exist in the modern Delta: a 96-97% loss of channels in this size class. This decrease has most likely reduced the population viability of native wildlife in these habitats by eliminating the large, widely distributed, and connected populations. The reduced extent of high-endemism habitats, such as vernal pools and alkali wetlands, may have significant consequences for biodiversity in the region (see Chapter 8). The effects of habitat loss, fragmentation, and degradation on marsh and riparian wildlife are discussed in Chapters 5 and 7. As a result of the diking of marshes, dendritic channel networks have been lost, with ecological consequences for native fish (see Chapter 4). The reduction of high-productivity marsh habitat has reduced the food resources available for fish and waterfowl (discussed in Chapter 4 and Chapter 6). In general, the scale-dependent effects of habitat loss on food resources are not well understood. Marsh production, from the marsh plain and the shallow, high-water-residence-time dendritic channels, was undoubtedly consumed and sequestered within the marsh, as well as being consumed by transient and edge wildlife, with some productivity ultimately being exported in one form or another to the broader estuarine and adjacent terrestrial ecosystems.[7]

Historically there was considerable geomorphic and hydrological heterogeneity within Delta habitats, creating diverse options for wildlife. This heterogeneity grew from the complex and variable hydrology, water and air temperature gradients, and differences in geomorphic setting, including topography and soils.[8] These differences manifested as diversity in plant communities and water chemistry, which provided a variety of options for wildlife. The riparian shrub habitats of the south Delta supported different species than the wide riparian gallery forests of the north Delta (see Chapter 7). Likewise, the dense tule marshes of the north Delta, willow-interspersed marshes of the central Delta, and complex marsh mosaics of the south Delta likely supported somewhat different communities of marsh wildlife (see Chapter 5 and **Chapter 6**). Yet some broadly distributed species with an ability to exploit diverse habitats, like Song Sparrows and Virginia Rails, were likely present across all these types of marsh, as large and diverse populations. Heterogeneity within habitats provided niche opportunities and increased habitat complexity, which is one way to create and maintain the genotypic and phenotypic diversity necessary for adaptation to change. Thus, heterogeneity supported the adaptation potential of wildlife and, in some cases, the development of alternative life-history strategies.[9] Heterogeneity within the Delta allowed different runs of Chinook salmon to exploit different resources at different times of year, supporting the diversity in salmon life-history strategies present today (see Chapter 4).

The modern Delta has lost connectivity within and among habitat types. Once-continuous populations of marsh species are now dispersed metapopulations or small, isolated populations at risk of extirpation. Riparian forests that once were unbroken corridors for terrestrial wildlife movement are now small, isolated, narrow patches often disconnected from the flooding that sustains them. Other habitat types in the Delta are also disconnected from one another, bounded by levees and separated by a matrix of agriculture. Approximately 1,770 km of levees exist in the modern Delta, separating channels and marshes from adjoining habitats. Historical flooding moved sediment, nutrients, and organisms between adjacent habitats, replenishing less productive areas

Damming a Delta slough.

Unknown ca. 1900, MS 229, Dyer Photograph Album, courtesy of Holt-Atherton Special Collections, University of the Pacific Library

on a regular basis and maintaining geomorphic structure. Loss of connectivity in the modern Delta disrupts these water and energy flows, impacting productivity[10] and resilience. Loss of habitat connectivity also reduces the viability of wildlife populations by restricting gene flow and limiting the ability of individuals and species to move conditions change.[11] One exception is that connections between large channels have increased over time as a result of channel cuts and dredging. The over connectivity of the channel network, and abundance of looped channels (combined with altered flow regimes) results in flow paths and chemical signals that are unpredictable for aquatic species.[12]

The quality of remaining habitats within the Delta has been degraded by a loss of complexity and the addition of anthropogenic stressors. The channels that now characterize the Delta are wider, straighter, deeper, and simpler than historical channels, and generally lack the fine-scale structure and micro-topography (e.g., from pools, vegetated banks, channel cut-offs, and backwaters) that once increased habitat value for aquatic wildlife. High nutrient loads and contaminants impair water quality and can reduce wildlife survival and reproductive success.[13] Invasive species have altered food-web dynamics, particularly the Asian clam, which reduces phytoplankton availability.[14] Introduced predatory fish, like bass and sunfish, directly compete with and prey upon native fish.[15] Wetland and upland habitats have also suffered the effects of introduced species such as *Arundo* and Himalayan blackberry, both of which can dramatically alter habitat structure and diversity. Grasslands along the edge of the Delta have been almost entirely converted from perennial grasses and forbs to non-native annual grasses (see Chapter 8).

Habitat types are now disconnected from the processes that created and sustained them. Rivers and sloughs are separated from their floodplains by artificial levees, so flood waters do not deliver the sediment and nutrients to adjacent lands. Most leveed agricultural land has subsided to well below sea level. Similarly, riparian forests are no longer inundated by the floods that maintained the natural levees they grow upon. Upland habitat types now occupy topographic lows. The naturally dynamic and seasonal hydrology of the Delta has been greatly simplified and constrained. Lakes, ponds and basins are now often disconnected from the larger channel network, and no longer fill with floodwaters during the winter and then drain over the summer. Instead, they have become perennial warm-water habitat that favors invasive fish.[16] Though not historically a delta of actively migrating meanders,[17] tidal channels have been deepened, widened, and straightened- their edges hardened- limiting their ability to adjust and respond to environmental changes. The rivers that feed the Delta have been almost uniformly dammed and their channels armored and leveed, simultaneously cutting off peak flows, reducing sediment supply, and altering seasonal hydrology.

These and other interruptions or constrictions of physical processes have contributed to the development of a brittle skeleton of the former Delta, pinned in place by roads and levees, and unable to benefit from the processes that created it. Thus, the changes in physical processes mirror the changes in habitat. Both have been so severely altered and reduced that the dominant features of the historical Delta – extensive marshes nourished with seasonal flooding and supporting vast wildlife populations – are no longer present. The Delta today is a network of deep, engineered channels within a matrix of leveed agriculture, supporting declining native wildlife and increasing invasive species populations.

The following pages describe overall change in habitats and the channel network. These changes are the easiest to quantify, given the available historical and contemporary datasets. Changes in habitat quality, habitat heterogeneity, and physical processes are often described qualitatively, since the datasets necessary for quantification are not available. These overarching analyses provide context for understanding the changes in ecological functions which are assessed in the subsequent chapters.

(top) Riprap and oaks on artificial levee, Lindsey Slough. (bottom) Dredges creating meander cuts on the San Joaquin.

Top: Erin Beller, SFEI; Bottom: Covella & Farichild ca. 1910, courtesy of Bank of Stockton Historical Photograph Collection

The extent of habitat type conversion has been extreme

The Delta has been converted from a marsh-dominated landscape to an agriculture-dominated landscape

The historical Delta was characterized by a complex and extensive marshland matrix. Broad corridors of riparian forest snaked down into the marsh along major rivers and distributaries. Seasonal wetlands and vernal pools lined the periphery of the north Delta. Willow thickets were interspersed throughout the tules in the central Delta. In the south Delta, tidal wetlands graded into non-tidal wetlands across a long, heterogeneous fluvial-tidal interface. While many of the shapes of these former features can still be identified in the contemporary Delta, habitat type conversion to agricultural and urban development has been extreme. Small remnants and restored (both purposeful and accidental) habitats can be seen scattered throughout the system.

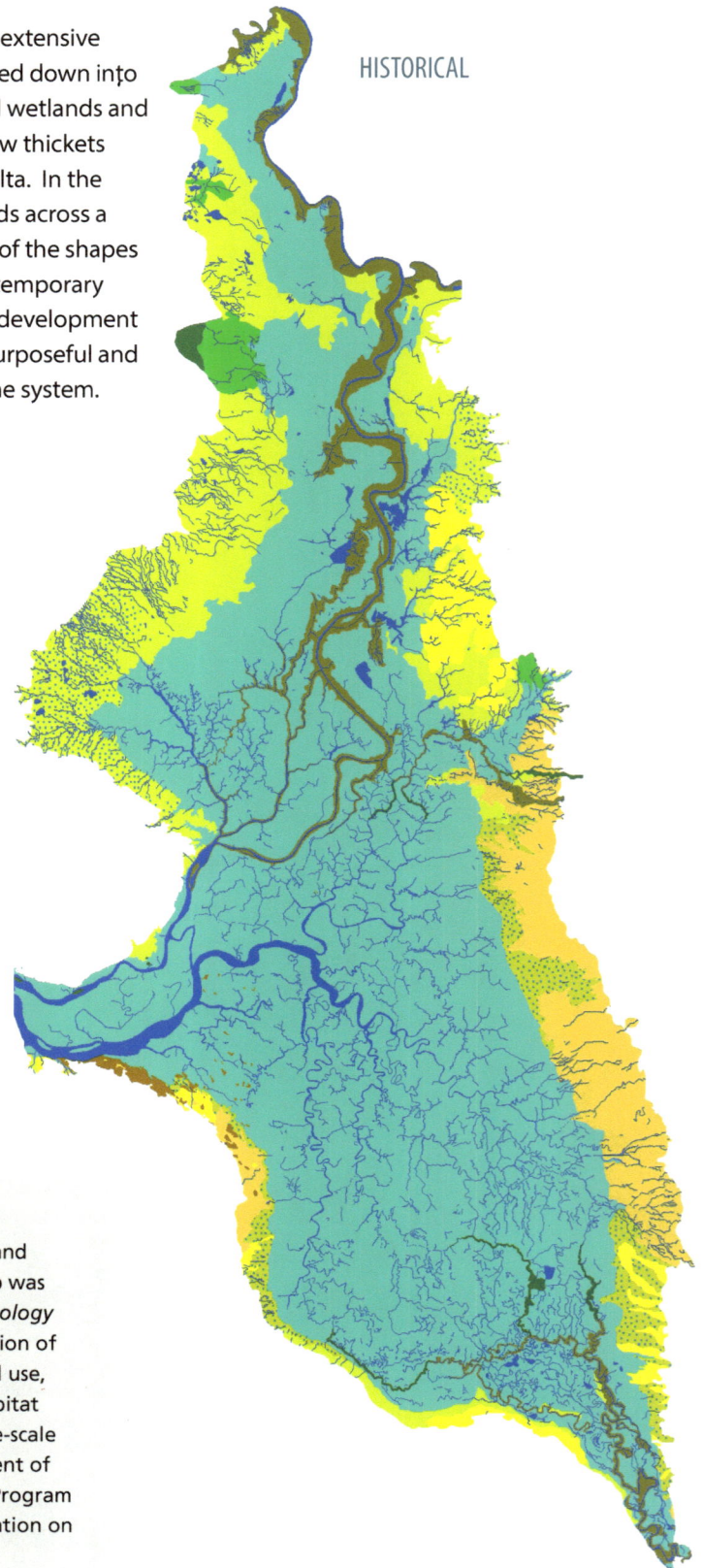

HISTORICAL

Delta habitat types, past (right) and present (far right). Historical habitat types and channels for the historical Delta ca. 1800 are shown to the right. Modern habitat type mapping ca. 2007 is shown to the far right.

Methods: Habitat type extent

Habitat type acreages were calculated from the historical and modern habitat type maps. The **historical habitat type map** was taken from the *Sacramento-San Joaquin Delta Historical Ecology Investigation.*[18] The **modern habitat type map** is a compilation of several spatial datasets detailing Delta vegetation and land use, with each vegetation type crosswalked to the historical habitat types. The majority of the modern map is derived from fine-scale vegetation mapping produced in 2007 by the CA Department of Fish and Wildlife's Vegetation Classification and Mapping Program (VegCAMP).[19] Please see Appendix A for additional information on developing the historical and modern habitat type layers.

MODERN

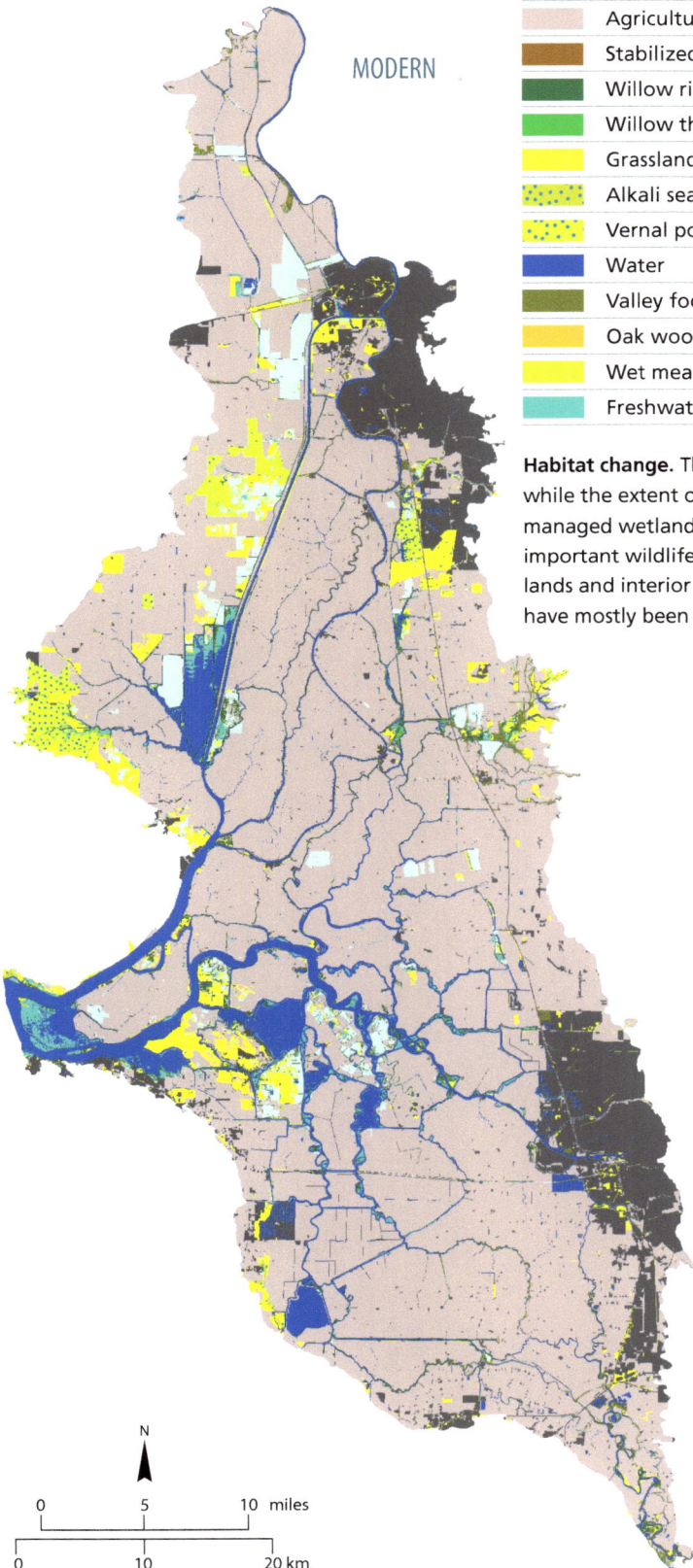

Habitat Type	Area (ha)		% Change
	Historical	Modern	
Managed wetlands	0	9,454	∞
Urban/Barren	0	35,517	∞
Agriculture/Non-native/Ruderal	0	216,085	∞
Stabilized interior dune veg.	1,032	4	-99
Willow riparian scrub/shrub	1,637	2,878	+76
Willow thicket	3,567	132	-96
Grassland	9,108	11,800	+30
Alkali seasonal wetland complex	9,193	238	-97
Vernal pool complex	11,262	3,007	-73
Water	13,772	26,530	+93
Valley foothill riparian	15,608	4,010	-74
Oak woodland/savanna	20,460	0	-100
Wet meadow/Seasonal wetland	37,561	2,445	-93
Freshwater emergent wetland	193,224	4,253	-98

Habitat change. The extent of wetland habitats has decreased in the modern Delta while the extent of open water and grasslands has increased. Agriculture and managed wetlands take up a large portion of the modern Delta and provide some important wildlife support but are not equivalent to historical habitats. Oak woodlands and interior dune scrub have mostly been eliminated.

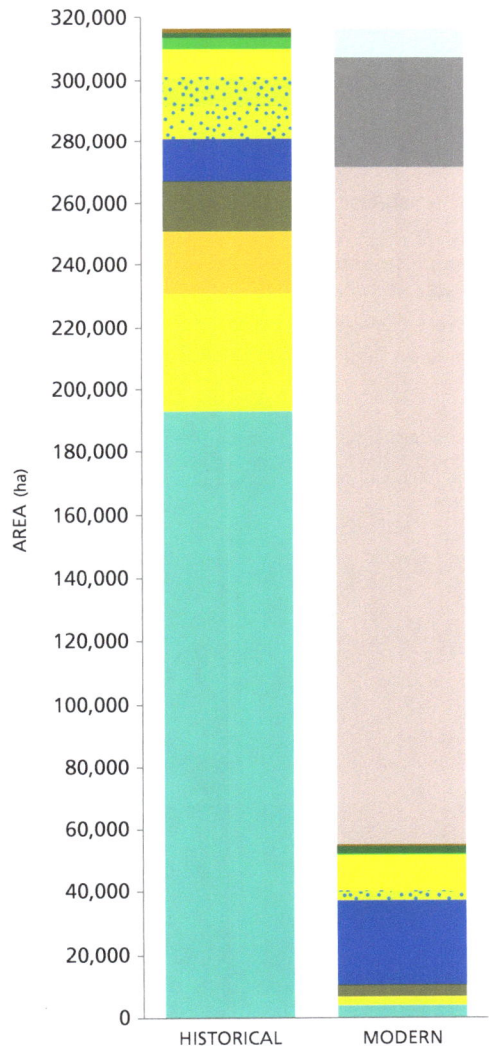

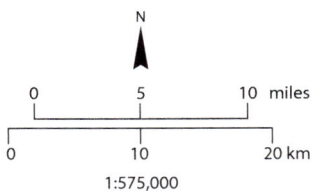

The variety of Delta habitats supported native wildlife diversity

The historical Delta supported a unique assemblage of species, contributing to the overall biodiversity of the region

Habitat diversity within the historical Delta contributed to overall species diversity. Much of the historical Delta was freshwater emergent marsh and aquatic habitat, which supported numerous species. Adjacent habitat types each supported distinct species assemblages and provided additional support to species that used the marsh and aquatic habitats. Many of the protected species found in the Delta today relied on varied habitat types historically (see far right).

Abundant resources from multiple habitat types and habitat adjacencies led to significant biodiversity in the historical Delta. There were also areas of importance to endemic and rare native species that disproportionately contributed to overall biodiversity. The introduction of invasive species has increased the total number of species in some areas, likely at the expense of native species diversity.[20]

Delta habitat types (right) and their affiliated species (far right). Each habitat type in the Delta supported specific suites of species, though several species used multiple habitat types for different phases of their lives. The species listed to the far right are BDCP Covered Species. Historical species-habitat type associations are based on modern species-habitat associations and life-history characteristics.[21]

Photo Credits (clockwise from top left): Dan Cox, USFWS; Steve Emmons, USFWS; Lee Eastman, USFWS; Brian Hansen, USFWS; Jon Katz and Joe Silveira, USFWS; Steve Martarano, USFWS

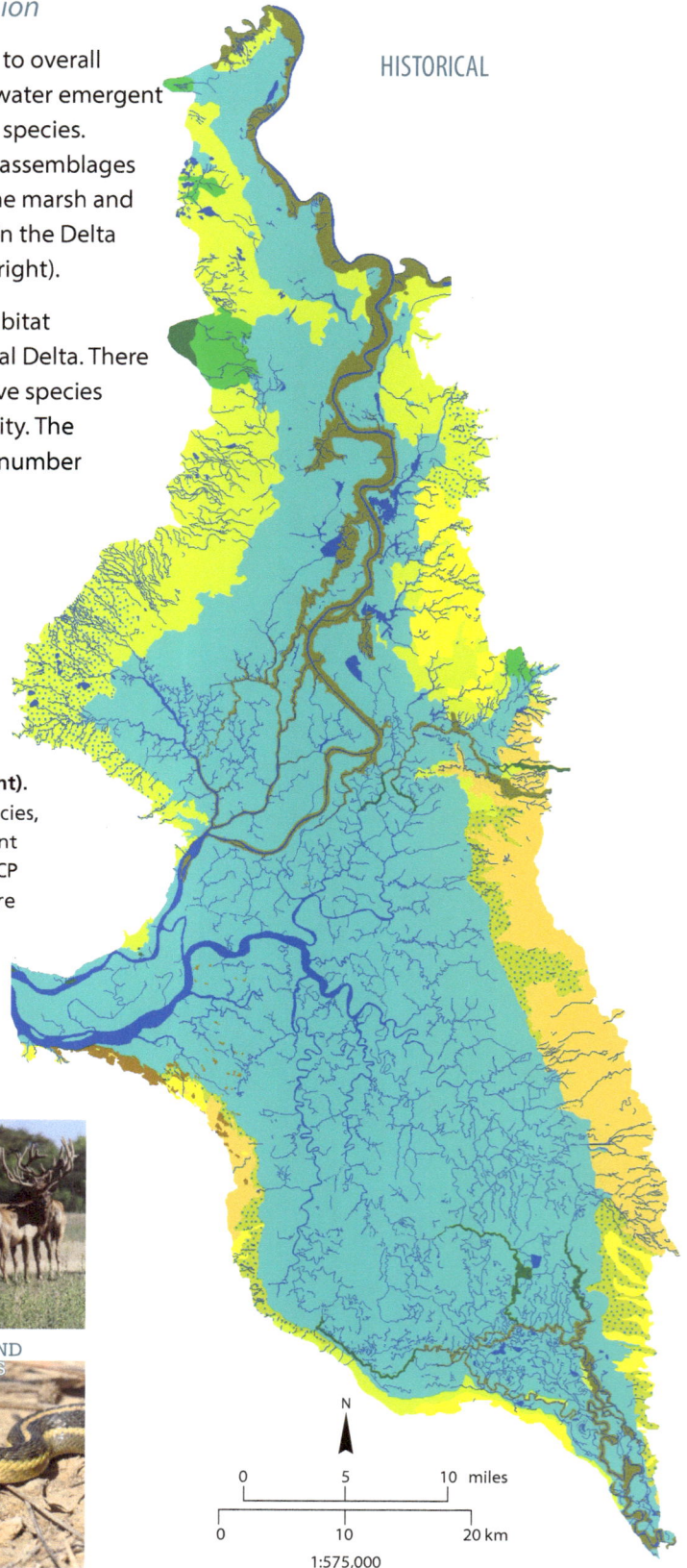

HISTORICAL

● FISH

● BIRDS

● MAMMALS

● PLANTS

● ARTHROPODS

● REPTILES AND AMPHIBIANS

N

| 0 | | 5 | | 10 miles |

| 0 | 10 | 20 km |

1:575,000

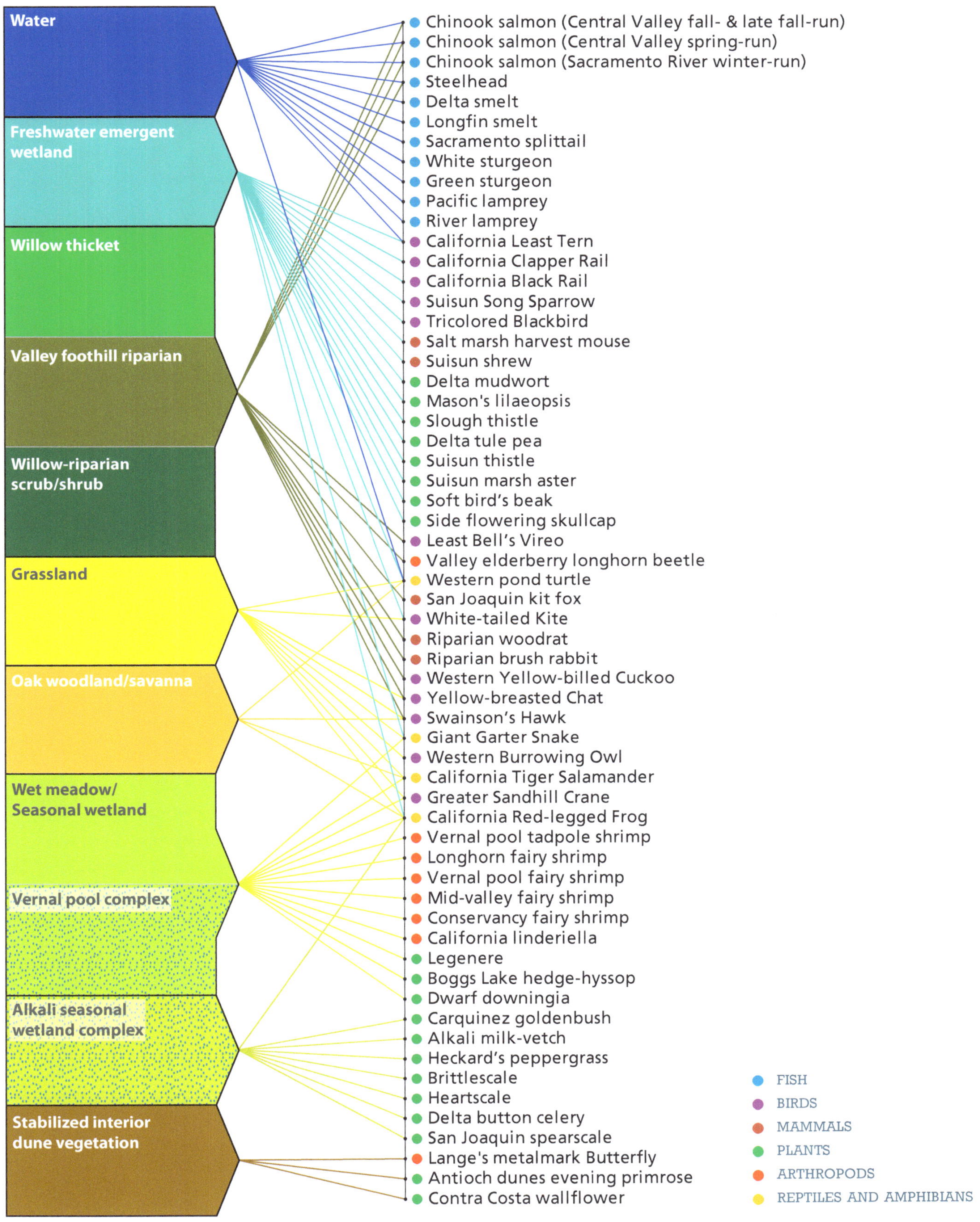
Water
Freshwater emergent wetland
Willow thicket
Valley foothill riparian
Willow-riparian scrub/shrub
Grassland
Oak woodland/savanna
Wet meadow/ Seasonal wetland
Vernal pool complex
Alkali seasonal wetland complex
Stabilized interior dune vegetation
Chinook salmon (Central Valley fall- & late fall-run)
Chinook salmon (Central Valley spring-run)
Chinook salmon (Sacramento River winter-run)
Steelhead
Delta smelt
Longfin smelt
Sacramento splittail
White sturgeon
Green sturgeon
Pacific lamprey
River lamprey
California Least Tern
California Clapper Rail
California Black Rail
Suisun Song Sparrow
Tricolored Blackbird
Salt marsh harvest mouse
Suisun shrew
Delta mudwort
Mason's lilaeopsis
Slough thistle
Delta tule pea
Suisun thistle
Suisun marsh aster
Soft bird's beak
Side flowering skullcap
Least Bell's Vireo
Valley elderberry longhorn beetle
Western pond turtle
San Joaquin kit fox
White-tailed Kite
Riparian woodrat
Riparian brush rabbit
Western Yellow-billed Cuckoo
Yellow-breasted Chat
Swainson's Hawk
Giant Garter Snake
Western Burrowing Owl
California Tiger Salamander
Greater Sandhill Crane
California Red-legged Frog
Vernal pool tadpole shrimp
Longhorn fairy shrimp
Vernal pool fairy shrimp
Mid-valley fairy shrimp
Conservancy fairy shrimp
California linderiella
Legenere
Boggs Lake hedge-hyssop
Dwarf downingia
Carquinez goldenbush
Alkali milk-vetch
Heckard's peppergrass
Brittlescale
Heartscale
Delta button celery
San Joaquin spearscale
Lange's metalmark Butterfly
Antioch dunes evening primrose
Contra Costa wallflower
FISH
BIRDS
MAMMALS
PLANTS
ARTHROPODS
REPTILES AND AMPHIBIANS

The Delta is a highly invaded system

Invasive species have altered the functions and quality of Delta habitats

The Delta has been inexorably altered by the introduction of numerous non-native species. These species have changed not only the community composition of Delta wildlife, with non-native species outnumbering native species in some instances, but have also affected the structure, functions, and processes the Delta can support. The alteration of physical processes and habitats in the Delta has undoubtedly facilitated some of these invasions, and the invasions themselves have further altered and degraded habitats within the Delta. The proliferation of non-native species within the Delta places considerable constraints on the extent to which restoration and other management actions can benefit native species. Non-native species affect the ability of the Delta to support native wildlife through several mechanisms, including habitat alteration, changes in food web structure, competition, and predation (see examples below).

ALTERED HABITAT STRUCTURE

Egeria (shown left) changes flow patterns and turbidity in shallow aquatic habitats. *Arundo* and Himalayan blackberry form dense thickets, impenetrable to some wildlife, in both marsh and riparian habitats.

CHANGED FOOD WEB STRUCTURE

The high filtration rate of the now abundant overbite clam has substantially reduced phytoplankton availability in the Delta. The invasion of the Delta by this clam is correlated with a stepwise decline in fish abundance (Pelagic Organism Decline).

INCREASED PRESSURE ON NATIVE SPECIES

Non-native fish introduced to the Delta for sport, including striped bass (shown left) and bluegill, compete with native species for limited resources. Non-native predators, including feral pets and nuisance species such as house mice and introduced rats, increase predation pressure on native wildlife.

Photo Credits: California Department of Boating and Waterways; Delta Stewardship Council; Dave Giordano, Cal Fish

Methods: Areas dominated by non-native and invasive plants

Individual polygons were marked as dominated by non-native/invasive vegetation if their specific alliance/association-level vegetation mapping unit featured a non-native (as defined by CalFlora)[22] or invasive species (as defined by the California Invasive Plant Council).[23] Where these fine scale classifications were not available, the non-native/invasive designation was determined based on group-level mapping units and best professional judgement. Areas with a habitat type of Agriculture/Non-native/Ruderal or Urban/Barren were classified by default as non-native. See pages 106-109 of Appendix A for a list of the mapping units classified as dominated by non-native/invasive vegetation. Invasive submerged aquatic vegetation is included in the mapping, but may be underrepresented depending on the year and season.

MODERN

Dominated by non-native vegetation
Dominated by invasive vegetation

PERCENT OF TOTAL HABITAT TYPE AREA

100%
90%
80%
70%
60%
50%
40%
30%
20%
10%
0%

AGRICULTURE/NON-NATIVE/RUDERAL
URBAN/BARREN
GRASSLAND
VERNAL POOL COMPLEX
WET MEADOW/SEASONAL WETLAND
WILLOW RIPARIAN SCRUB/SHRUB
MANAGED WETLAND
VALLEY FOOTHILL RIPARIAN
FRESHWATER EMERGENT WETLAND
WATER
ALKALI SEASONAL WETLAND COMPLEX
INTERIOR DUNE SCRUB
WILLOW THICKET

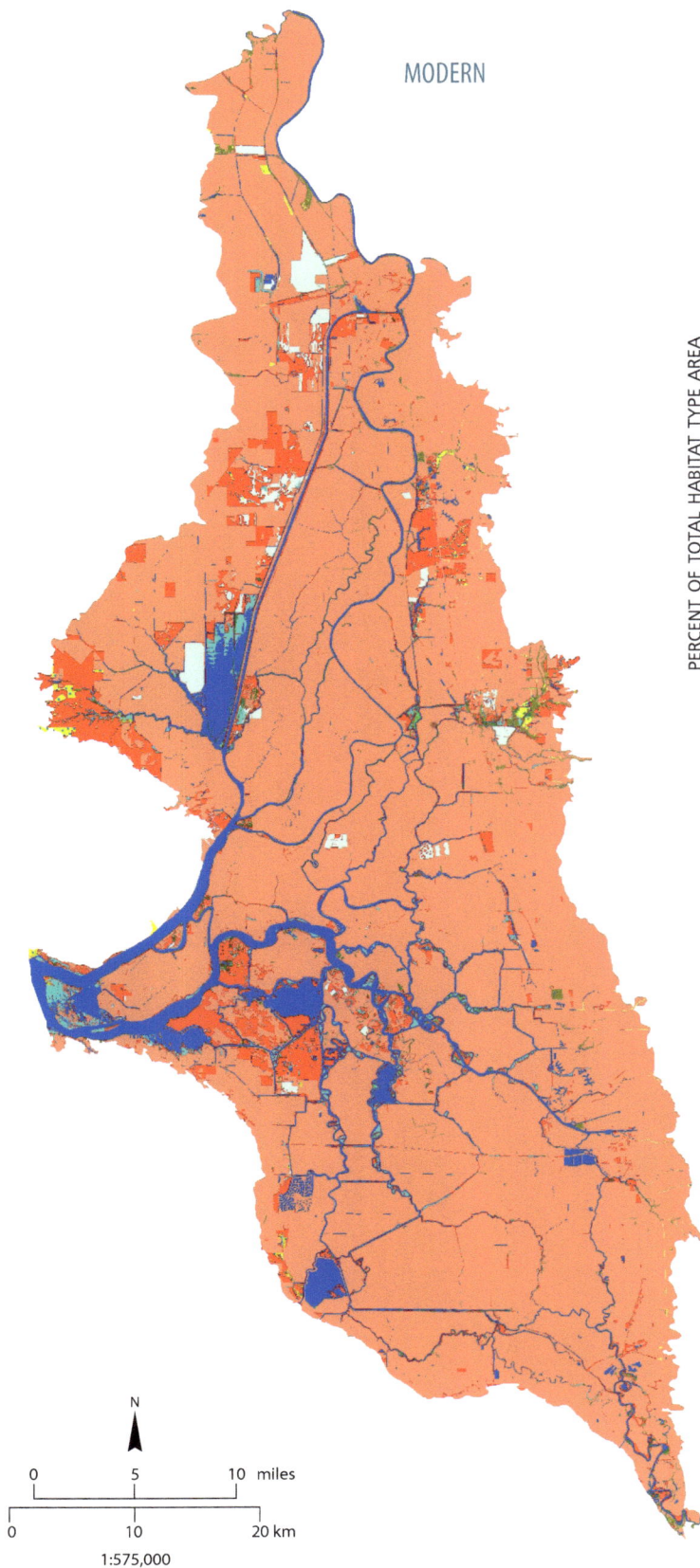

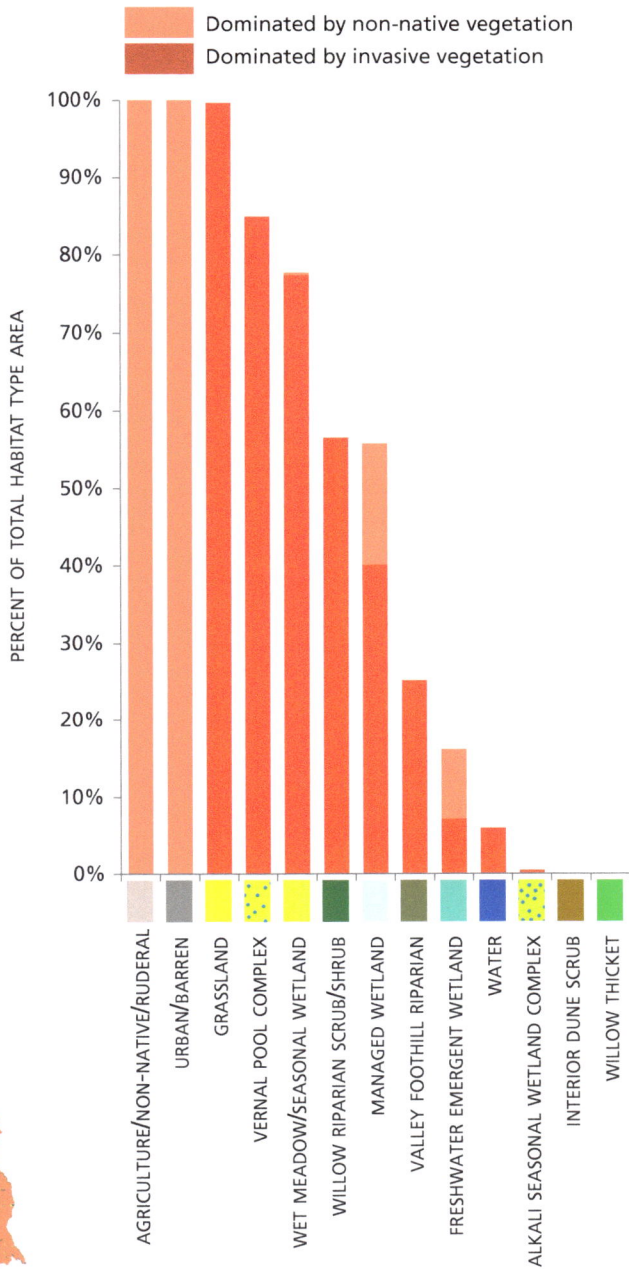

Areas dominated by non-native or invasive plants. The map **(left)** shows areas where non-native (in pink) and invasive (in red) plants are dominant over or co-dominant with native vegetation. The chart **(above)** quantifies the total proportion of each habitat type dominated by non-native/invasive vegetation.

N

0   5   10 miles

0   10   20 km

1:575,000

The Delta's channel network and lakes have been fundamentally altered

Some dominant native aquatic habitat types have been nearly eliminated, while other novel types have been created

The aquatic habitats of the Delta have been changed in several ways, all of which are significant to ecological functions. New channels have been dug for shipping, creating new, often straight and leveed waterways (A). Perhaps most severe has been the filling and elimination of the branching dendritic tidal channels that wove through the marshes. These channels were often narrow and shallow with high residence times providing important habitat for fish species (B). Several previously farmed and diked islands in the Delta (such as Sherman Island, Franks Tract, and Liberty Island) have drowned or are in the process of drowning due to subsidence and levee failure, leaving in their wake more open and deep water than existed in the historical Delta (C). While much of the Sacramento River has maintained a consistent width due to its natural levees, many reaches in the central Delta have been widened (A, C). Throughout the Delta, existing channels have been hyper-connected through channel cuts and meander cut-offs (D). This lowers residence times and often increases average velocities, thereby providing less nutrients, shelter, and habitat complexity for aquatic species. Finally, while the San Joaquin River has continued to migrate (more than the Sacramento), off-channel aquatic habitat such as floodplain and oxbow lakes and distributary channels has been filled (E).

Reconfiguration of the aquatic landscape (historical and modern) (right). Historical aquatic habitats (in yellow) are overlaid with modern aquatic habitats (in blue). Areas where past and present aquatic habitat overlap are displayed as green.

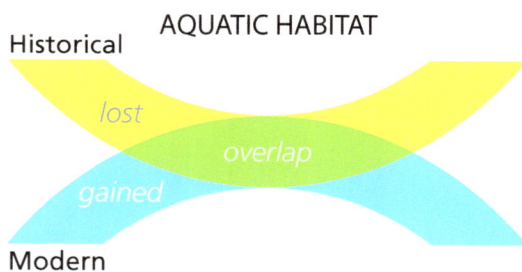

AQUATIC HABITAT

Historical

lost

overlap

gained

Modern

Methods: Changes in channel planform

Determining areas of aquatic habitat that have been **lost, gained, or have not changed (overlap)** was achieved by intersecting areas classified as aquatic habitat in the historical and modern habitat types layers.

Channel width (top of facing page) was calculated at 100 m intervals by casting transects perpendicular to the channel centerline, clipping the transects to the banks of the channel, and subtracting any portion of the width associated with in channel islands.

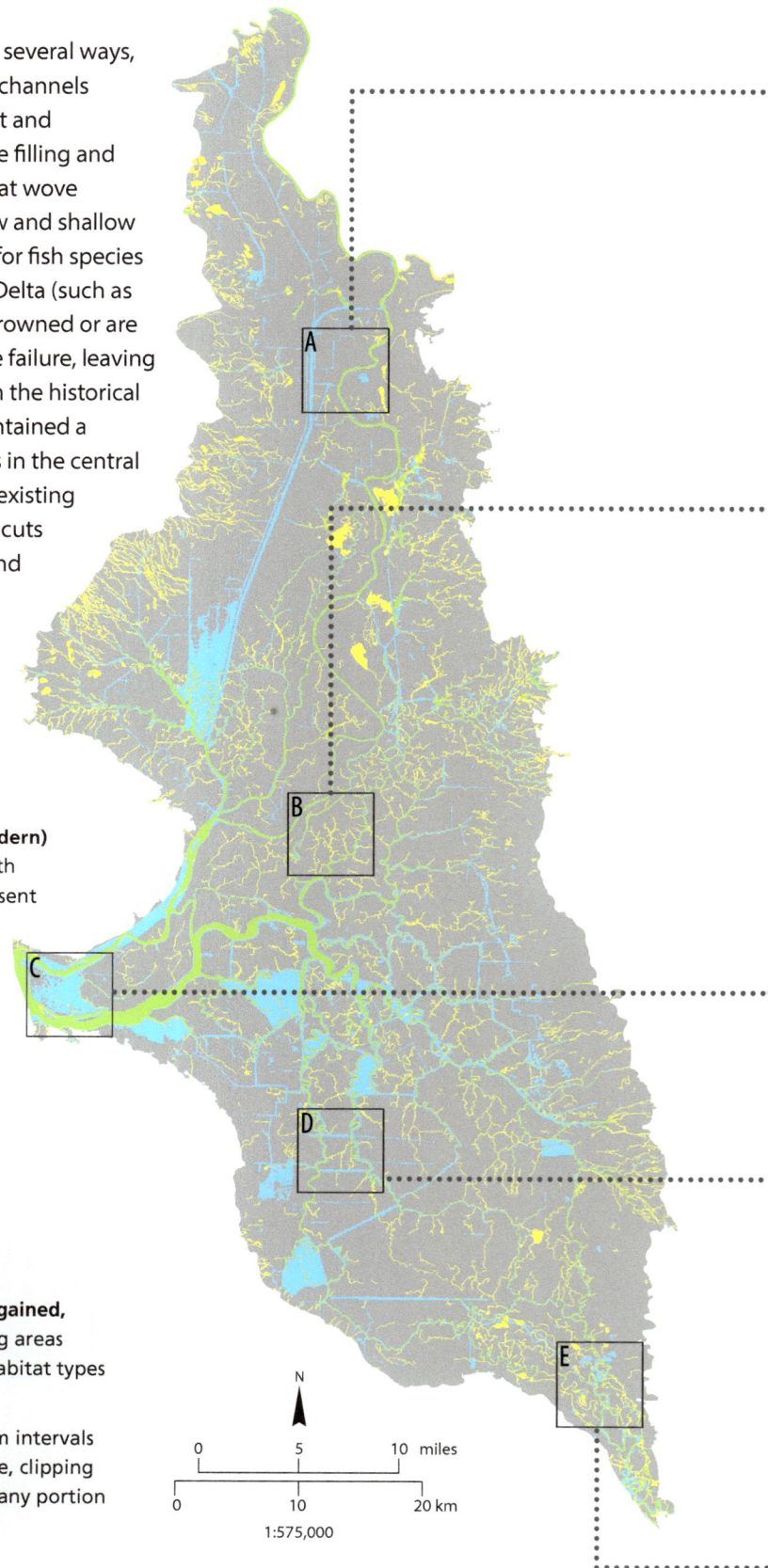

N

| 0 | 5 | 10 miles |

| 0 | 10 | 20 km |

1:575,000

A

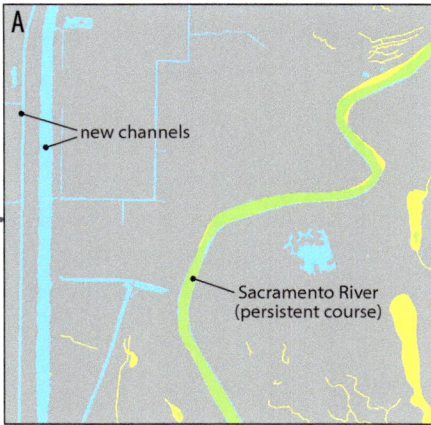

Confined by large natural levees, the course and width of the Sacramento River (here in green) is largely unchanged, but new straight channels have been created (blue).

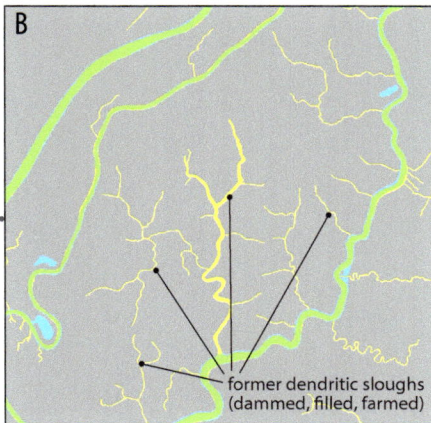

new channels

Sacramento River (persistent course)

The loss of narrow channels and increase in wide channels (above). The most significant change when comparing channel widths is seen in the lowest width category (0-15 m). The length of narrow channels in the Delta has decreased by two orders of magnitude, effectively eliminating more than 5,000 km of channels. The length of channels in the wider size classes (between 100-1,000 m) has essentially doubled.

B

The smaller, dendritic, dead end sloughs of the Delta (here in yellow) have almost all been diked and filled.

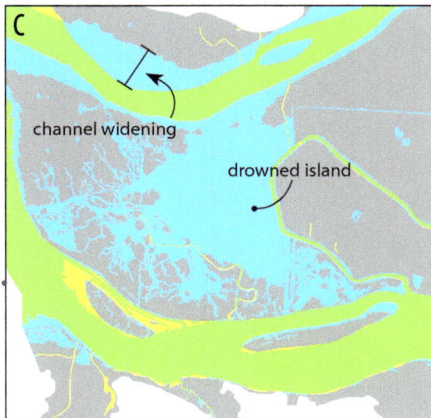

former dendritic sloughs (dammed, filled, farmed)

Several types of reconfiguration of aquatic habitats (A-E, left and below). Small channels have been diked and filled. Large channels have been straightened, leveed, and artificially connected. Since the geometry of the Delta largely controls the dispersion and trapping of tidal waters,[25] these changes have likely had significant impacts on key physical processes and gradients (e.g., tidal flows, sediment transport and deposition, salinity transport, water residence time, water temperature, terrestrial linkages).

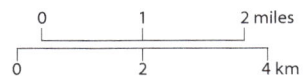

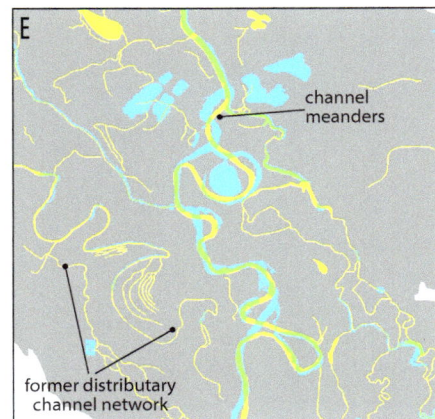

C

Some channels, like the lower Sacramento, have been substantially widened. Levee breaches flood subsided islands, creating extensive new areas of open water.

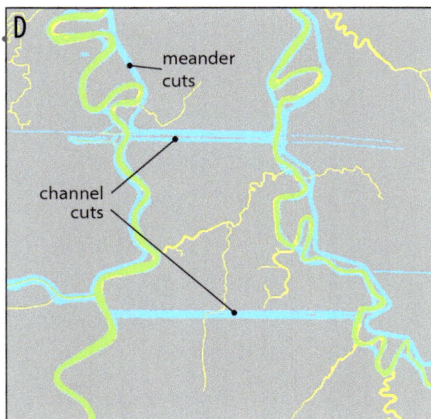

channel widening

drowned island

D

Meander cuts (between bends in a channel) and channel cuts (between separate sloughs) effectively straighten and short-circuit tidal channel networks.

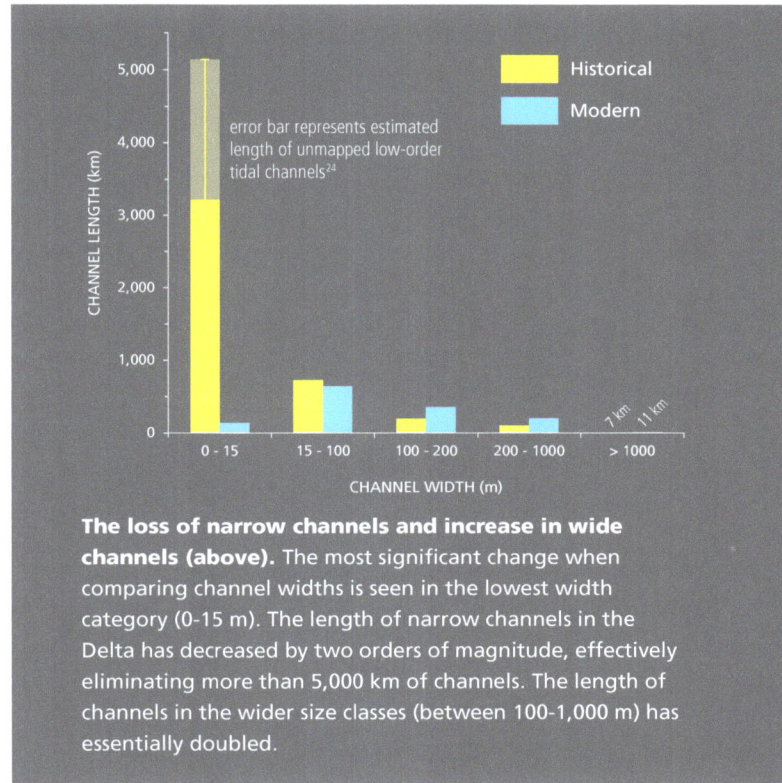

meander cuts

channel cuts

E

The main course of the San Joaquin has meandered over time. Its smaller floodplain channels have been mostly filled.

channel meanders

former distributary channel network

There is twice as much tidal shallow-water habitat in the Delta today as there was historically

Shallow dendritic channels and lakes have been exchanged for novel flooded island habitats

Existing tidal shallow-water habitat types are different than they were historically. The majority of areas <2 m deep today is part of large open water expanses rather than the historical small marsh channels. Shallow channel habitat is now found mainly along the edges of larger channels and flooded islands, adjacent to deep water. In fact, because of the widening of large channels, construction of new channels, and accidental flooding of subsided islands, there is more tidal aquatic habitat today in all depth classes, despite the near total loss of small marsh channel networks.

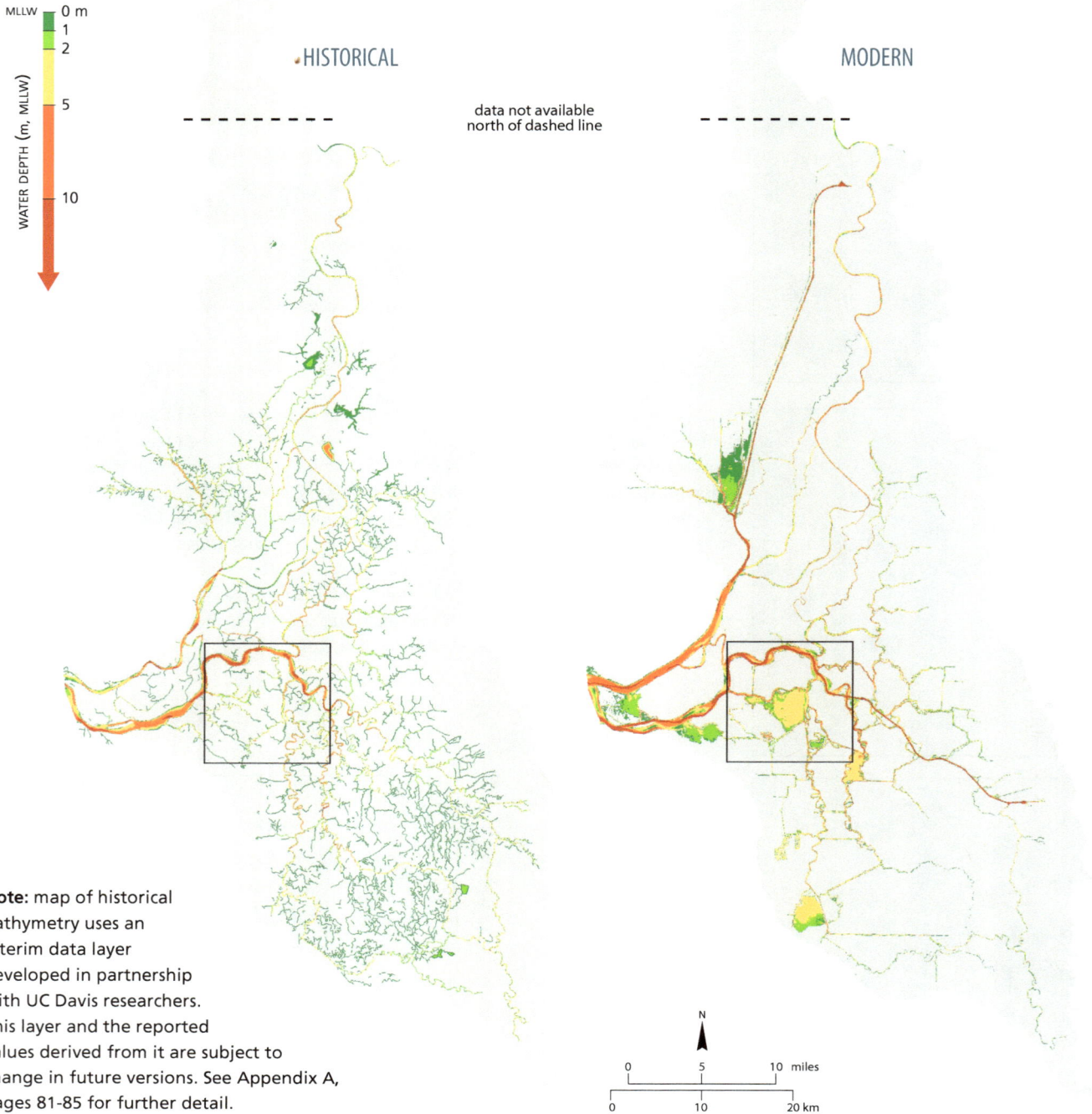

MLLW
0 m
1
2

5

WATER DEPTH (m, MLLW)

10

HISTORICAL

MODERN

data not available north of dashed line

Note: map of historical bathymetry uses an interim data layer developed in partnership with UC Davis researchers. This layer and the reported values derived from it are subject to change in future versions. See Appendix A, pages 81-85 for further detail.

N

| 0 | 5 | 10 | miles |

| 0 | 10 | 20 km |

Near elimination of shallow channels; near doubling of deep channels. In the historical Delta, the vast majority of tidal channels (by length) were shallow (0-1 m). Nearly 1,500 km of low order dendritic channels in tidal channel networks likely had high water residence times and low velocities (see green channels below). These channels were almost completely eliminated during reclamation of the marshes. Large channels (the deepest classes: 5-10 m and >10 m) have greatly increased in extent, likely due to dredging and other modifications.

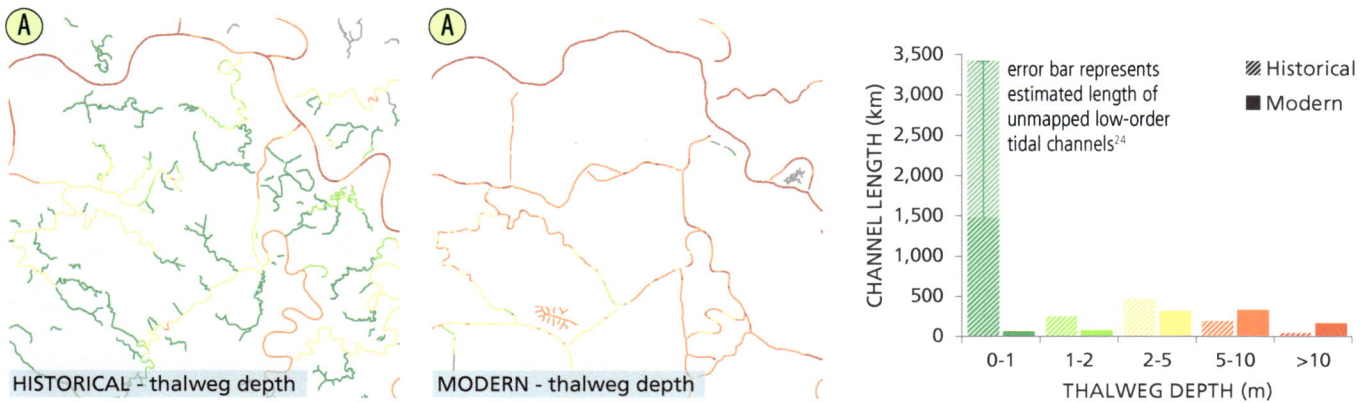

HISTORICAL - thalweg depth

MODERN - thalweg depth

error bar represents estimated length of unmapped low-order tidal channels[24]

Loss of lakes; creation of novel flooded islands. There has been an overall increase in the area of aquatic habitat and, in particular, shallow water in the modern Delta. Shallow-water habitat is now mainly found in flooded islands, and widened channels. Yet shallow flooded islands are not equivalent to the historical lakes, with different hydrologic patterns, they have been largely overtaken by invasive submerged aquatic vegetation and invasive species. The shallow water on the edges of large, deep channels in the modern Delta may provide refuge for some fish, but these areas also harbor invasive aquatic species and likely do not provide the same benefits to fish as small, dendritic channels.

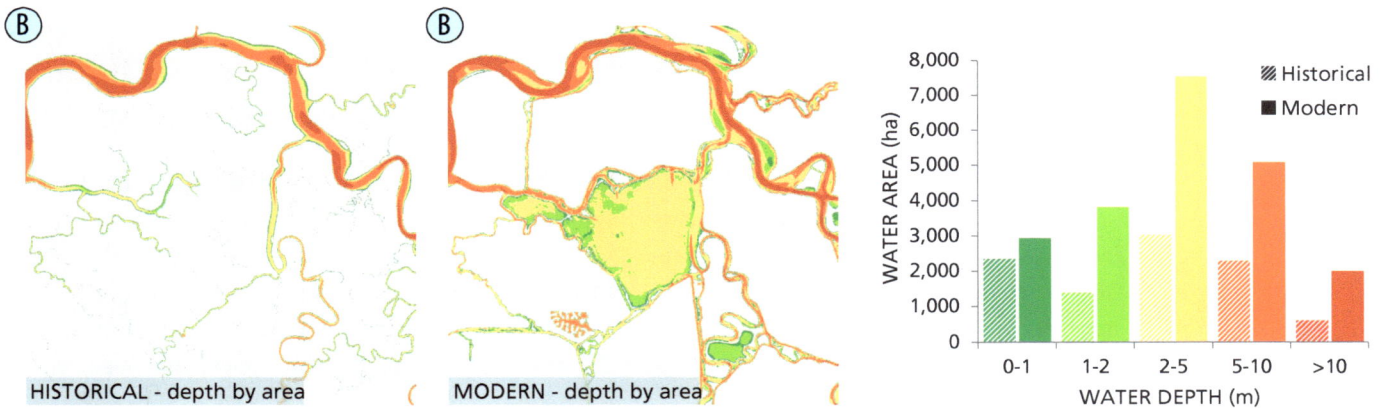

HISTORICAL - depth by area

MODERN - depth by area

Methods: Two ways to summarize extent of aquatic habitats of different depths

(A) **By length:** we summed the linear extent of channels based on their thalweg depth (taken at ~100 m intervals). The thalweg is the deepest part of the channel. For all analyses on this page, depth was measured from an approximate mean lower-low water (MLLW) elevation.[26]

(B) **By area:** We summed the areal extent of aquatic habitat at each depth class (water column depth).

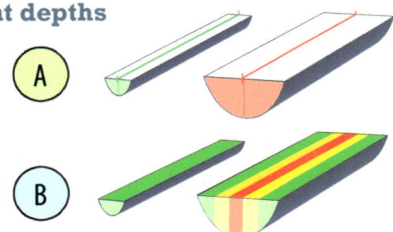

4. Life-History Support for Resident and Migratory Fish

Aquatic habitats in the historical Delta were complex and dynamic, providing many resources and opportunities for native fish. The rivers and sloughs that wove through the Delta displayed wide variation in width, depth, and sinuosity, creating heterogeneity in local hydraulics, residence time, and water chemistry. These characteristics provided diverse food resources and refuge for fish populations.[1] Historically, large channels flanked by riparian forest or marshes served as migration corridors for fish and provided resting places and refuge in undercut banks, deep pools, and inner bends.[2] Off-channel ponds and lakes were characterized by extensive shallow, slow-moving waters, which facilitated primary and secondary production for rearing populations.[3] Dendritic tidal channels that terminated in the marsh were backwaters with high residence times, and were characterized by temperature gradients beneficial for juvenile fish.[4] Delta channels were hydrologically connected to floodplains and marshes, and expanded in times of high water. Seasonally inundated floodplains offered a rich source of food and habitat for rearing and spawning.[5] Tidal flooding allowed fish access to the vegetated marsh and facilitated exchange of nutrients and organic matter between wetlands and open water habitats.[6] While the position of the large tidal channels, natural levees, and lakes in the Delta remained relatively unchanged from year to year, the seasonal and interannual variability in hydrology and weather created a complex and ever-changing portfolio of aquatic habitat available to fish through time.[7]

Historically the Delta supported an abundant and diverse fish community that included several species of anadromous fish and numerous endemic species, including two locally endemic species of smelt. The fish community included both freshwater stenohaline (narrow salinity tolerance) and euryhaline (broad salinity tolerance) species.[8] Fish confined to freshwater included hardhead, hitch, roach, Sacramento pikeminnow, and Sacramento sucker. These species also inhabited the tributaries that fed into the Delta.[9] Freshwater euryhaline species, associated primarily with freshwater but more tolerant of brackish conditions, included tule perch, Sacramento splittail, and both the longfin and delta smelt.[10] These species were found in Suisun Bay as well as the Delta. Euryhaline marine species such as staghorn sculpin and starry flounder were commonly associated with higher salinities but were able to tolerate freshwater conditions in the Delta.[11] Large numbers of anadromous fish passed through the Delta historically, taking advantage of the productive and protected Delta environment while migrating from freshwater to the ocean and back. These species included the Pacific and river lamprey, green and white sturgeon, Chinook salmon, and steelhead.[12] Chinook salmon were particularly abundant in the Delta, with four distinct runs and an estimated overall population of 1-2 million spawners per

year.[13] Many of the fish species that occupied the Delta were adapted to slower moving shallow waters and floodplains (habitats that have been largely eliminated in the modern Delta); these include the Sacramento perch (extirpated), thicktail chub (extinct), hitch, Sacramento blackfish, and Sacramento splittail.[14] Freshwater conditions predominated throughout the Delta, though high tides late in the season and during times of drought occasionally brought brackish water to the Delta mouth.[15]

Interpreting how the historical Delta supported fish is challenging because the current understanding of their natural history and ecology is based on their use of a heavily altered modern landscape. This difficulty is compounded by the dynamic nature of these aquatic habitats, which experienced tremendous temporal variability in the past. However, we can take a landscape-scale approach to understanding how the Delta historically supported fish and other aquatic wildlife. Within aquatic systems, as in terrestrial systems, different areas provide different habitat qualities, and boundaries between those areas affect the connectivity between them. These interactions take place at multiple scales.[16] Using this landscape-scale approach several aspects of the historical Delta stand out as particularly important for fish: (1) habitat heterogeneity, (2) presence of high-productivity habitats, and (3) connectivity among habitats.

Aquatic habitats were heterogeneous at multiple scales, providing support to wildlife at the individual, species, and community levels. Small-scale heterogeneity allowed individuals to escape unsuitable conditions. For example, channels, swales and microtopography on floodplains reduced stranding risk for rearing Chinook and splittail, while pockets of slow moving water, such as along inner undercut banks and submerged trees, allowed tule perch to occupy otherwise fast-flowing channels.[17] Large-scale heterogeneity allowed species to occupy different niches, preferentially occupying different positions along salinity, temperature, and turbidity gradients. While species such as thicktail chub may have been specifically adapted to slow-moving backwaters and lakes, species such as Sacramento splittail were able to take advantage of floodplain habitats, using these areas to spawn.[18] The heterogeneity of aquatic habitats allowed some species to develop multiple life-history strategies, each likely to be favored in different years and under different conditions. Chinook salmon, for example, exhibited a wide range of variability in the timing and location of spawning and rearing. This diversity in life-history strategies likely stabilized the population via portfolio effects, increasing resilience because different segments of the population were less likely to experience declines at the same time.[19]

Resident and migratory fish (below). Left to right: Sacramento splittail; Chinook salmon; Chinook salmon; tule perch.

Photo Credits: Unknown, USFWS; Dan Cox, USFWS; Blaine Bellerud, NOAA; Unknown, UC Davis

The Delta had several types of high-productivity habitats that supported the base of the food web. Within the water column, shallow water depths and high residence times likely supported high densities of phytoplankton.[20] Dendritic channels that terminated in the marsh and other backwater areas may have been particularly important in this regard.[21] Within open water habitats such as lakes, submerged and floating aquatic vegetation supported high densities of invertebrates that were important food sources for fish.[22] Periodically inundated marshes and floodplains contributed organic matter to fuel the food web. In the modern San Francisco Bay Delta Estuary, fish food webs are dependent upon autochthonous marsh materials,[23] and this dependence was likely even greater historically when more marsh habitat was available.[24] Delta fish likely varied their diet seasonally to take advantage of shifts in prey availability, while maintaining minimal dietary overlap among species, as has been observed in native fish in the modern Delta.[25] This ability to take advantage of diverse and dynamic food resources would have been beneficial to the fish community in the historical Delta.

Wetlands, including floodplains, were connected to aquatic habitats by regular, unimpeded flooding from tides, precipitation, and snowmelt. Water moved slowly through vegetated landscapes, allowing exchange between the channels and wetlands to occur and providing variation in water depths and velocities.[26] The pattern of wetland flooding, with pulses of inundation and slower recession, allowed fish to take advantage of these habitats while still being able to pass back into the river channels once floodplains began to dry. Floodplains were inundated for both short and long durations, providing temporally variable benefits to fish.[27] Connections to off-channel habitats affected water chemistry within the channels themselves.[28] Organic matter contributed by marshes would have increased turbidity.[29] Exchange of primary productivity and export of invertebrates would have affected the food web.[30] Riparian trees and shrubs contributed woody debris that altered flows, channel dynamics, and sedimentation processes, particularly in the south Delta.[31]

Floodplains were critical for fish migration, spawning, and rearing. Floodplains served as important rearing habitat for several species of resident and migratory fish.[32] Floodplain habitats provided fish with refuge from predation as well as from energetic demands and physiological stressors. These habitats had high turbidity and increased the extent of shallow-water habitat where certain species could hide.[33] The increased foraging space provided by floodplains may have reduced competition and the likelihood of encountering certain predators.[34] Native fish may have been vulnerable to predation by abundant birds, but this additional risk was likely offset by for increased growth on the floodplain and reduced predation risk later in the ocean.[35] Estuarine rearing in marshes and floodplains is important to Chinook salmon because it can reduce size-dependent mortality upon ocean entry by increasing the variation in the size and timing at which individuals reach the ocean.[36]

In the modern Delta, aquatic habitats are characterized by wider, deeper, straighter channels that are leveed off from adjacent habitats. There is now much less seasonal and spatial variation in hydrology and habitat. Connectivity between large channels has increased through connecting canals, meander cutoffs, cross-levees, and dredged and widened channels. This has homogenized conditions (e.g., salinity, temperature, nutrients, and flows) and altered tidal and flood routing through the Delta. The modern channel network no longer predictably leads to fluvial sources or dendritic channels, making the Delta a much less coherent landscape for native fish to navigate.[37] Channel systems with coherent gradients allowed fish in the historical Delta to position themselves where conditions were most suitable, despite the dynamic nature of these conditions. Delta smelt, for example, track the low salinity zone as it moves upstream and downstream seasonally. These fish use vertical migration and other behavioral adaptations to stay in favorable areas.[38] Native fish key in on changes in flow, water temperature, and turbidity to cue their movement.[39] Furthermore, where once fish could predictably travel a short distance between one habitat (e.g., a large fluvial channel with high velocities and low residence time) and another quite different one (e.g., a small marsh channel with low velocities and high residence time), now these distances are much greater, and the path to get from one habitat to another is much less predictable.[40]

Most of the slow-water habitat, highly productive floodplains, and marsh-influenced habitats in which Delta fish species evolved are lost. The loss of wetlands, development of artificial levees, and the increase in the size and connectedness of channels has increased the speed at which water moves through the Delta. Most of the channels in the Delta today are lined by steep artificial levees that isolate the channels from adjacent habitats, and much of the habitat that was once marsh has been converted to agriculture. Flooding occurs, though in very limited areas, and is predominantly short-duration. Between 1935 and 1995, for instance, the frequency with which the Yolo Bypass experienced at least seven days of overflow in the spring decreased from ~80% of years to ~20% of years.[41] While remnants of several lakes persist, today most of the large areas of open water in the Delta are drowned islands. These deep water habitats, primarily in the central Delta, did not have functional equivalents in the historical Delta.[42]

The modern Delta is characterized by a suite of threats not faced by Delta fish communities historically. Highly managed hydrology, including diversions and pumps, alters directional flows often entraining fish.[43] Agricultural runoff and water discharges impact water quality.[44] In addition, introduced invasive species have restructured food webs, altered habitats, and directly outcompete native fish. The invasive *Corbicula* clam has dramatically reduced planktonic food resources available to fish.[45] Invasive submerged aquatic vegetation (SAV) species, such as Brazilian waterweed and water hyacinth, provide different structure and reach higher densities than native SAV species, and thus are not functionally equivalent.[46] Invasive SAV species provide habitat for non-native predatory fish and support invertebrates that are less favored in the diets of native fish species.[47]

The Delta fish community is now dominated by non-native species including sunfish, bass, catfish, and common carp.[48] Native species are generally associated with higher river flows and lower temperatures, although a few non-natives, including striped bass, white catfish, channel catfish, and American shad are also associated with high flows.[49] While floodplain inundation is critical for native fish migration, breeding, and rearing, floodplains are currently heavily used by non-native species.[50] However, native fish, adapted to the Delta's flood cycle, have been found to spawn and leave the Cosumnes floodplain earlier than non-native fish (thus avoiding stranding), and may be able to quickly take advantage of newly flooded habitats.[51] Food limitation in the modern Delta likely intensifies competition with non-native fish, as well as non-native predation on natives.[52]

The future of threatened fish species is uncertain and threats and stressors may continue to worsen. Restoration of habitat for native fish is difficult. Competing water interests make it challenging to re-establish historical flows that favor native fish, and improvements to water quality and habitat will likely favor non-native fish to some degree. Marsh and floodplain restoration have the potential to preferentially help native fish, though restoration would need to be implemented on a large scale to increase the likelihood of success due to the large variability in fish response to restoration activities.[53]

Fish likely benefited from dynamically inundated landscapes

Most of the temporarily flooded habitat available to fish has been lost in the modern Delta

By comparing the past and present, it is apparent that the Delta has shifted from a mosaic of subtidal, tidal, and seasonally or episodically flooded habitats to a landscape where most of the aquatic habitat is permanently subtidal. Historically, fish utilized abundant periodically available habitat for spawning, rearing, additional food resources, and refuge from predators. Specific floodplain-associated species in the Delta included Sacramento perch, thicktail chub, Sacramento splittail, and juvenile Chinook salmon.[54] Today, likely in part due to habitat losses, two of these species can no longer be found in the Delta—Sacramento perch are locally extirpated and thicktail chub are globally extinct.[55]

Although all types of inundation have decreased in extent over time, altered flow regimes, artificial levees, and drainage systems have effectively eliminated the seasonal long-duration flooding that persisted for months at a time in the historical Delta. Contemporary inundation associated with the Yolo Bypass and Cosumnes River floodplain is more akin to the shallow, seasonal short-term flooding that was common to the seasonal wetlands of the historical Delta.[56] This has important consequences for species like Sacramento splittail whose life-history strategies require longer periods of sustained inundation (and potentially enables alternate rearing strategies for juvenile salmon).[57]

Approximate maximum extent and type of inundation in the historical (right) and modern (far right) Delta. While the extent of perennial open water features has increased over time, areas that experience tidal inundation, seasonal short-term flooding, and seasonal long-duration flooding (defined at far right) have all decreased in extent (by 144,000, 40,000, and 59,000 ha, respectively).

Photo Credits: McCurry, courtesy of California History Room, California State Library, Sacramento; Yolodave, Wikipedia Commons

Methods: Type and extent of flooding

For the historical Delta, areas regularly subjected to inundation were derived from the map of historical habitat types, which were defined in part, by their typical hydrology.[58] Areas mapped as tidal freshwater emergent wetland, for instance, were classified in the inundation analysis as areas of "tidal inundation." See Appendix A for our complete methodology.

Since the modern habitat type dataset does not distinguish between tidal and non-tidal freshwater emergent wetland, a proxy was used to define the areas that currently experience tidal inundation. Specifically, areas were assigned the "tidal inundation" classification if they were mapped as freshwater emergent wetland, were adjacent to open water, and fell within the historical extent of tidal marsh. Other areas of inundation were identified, mapped, and classified after conducting a literature search and consulting with regional experts.

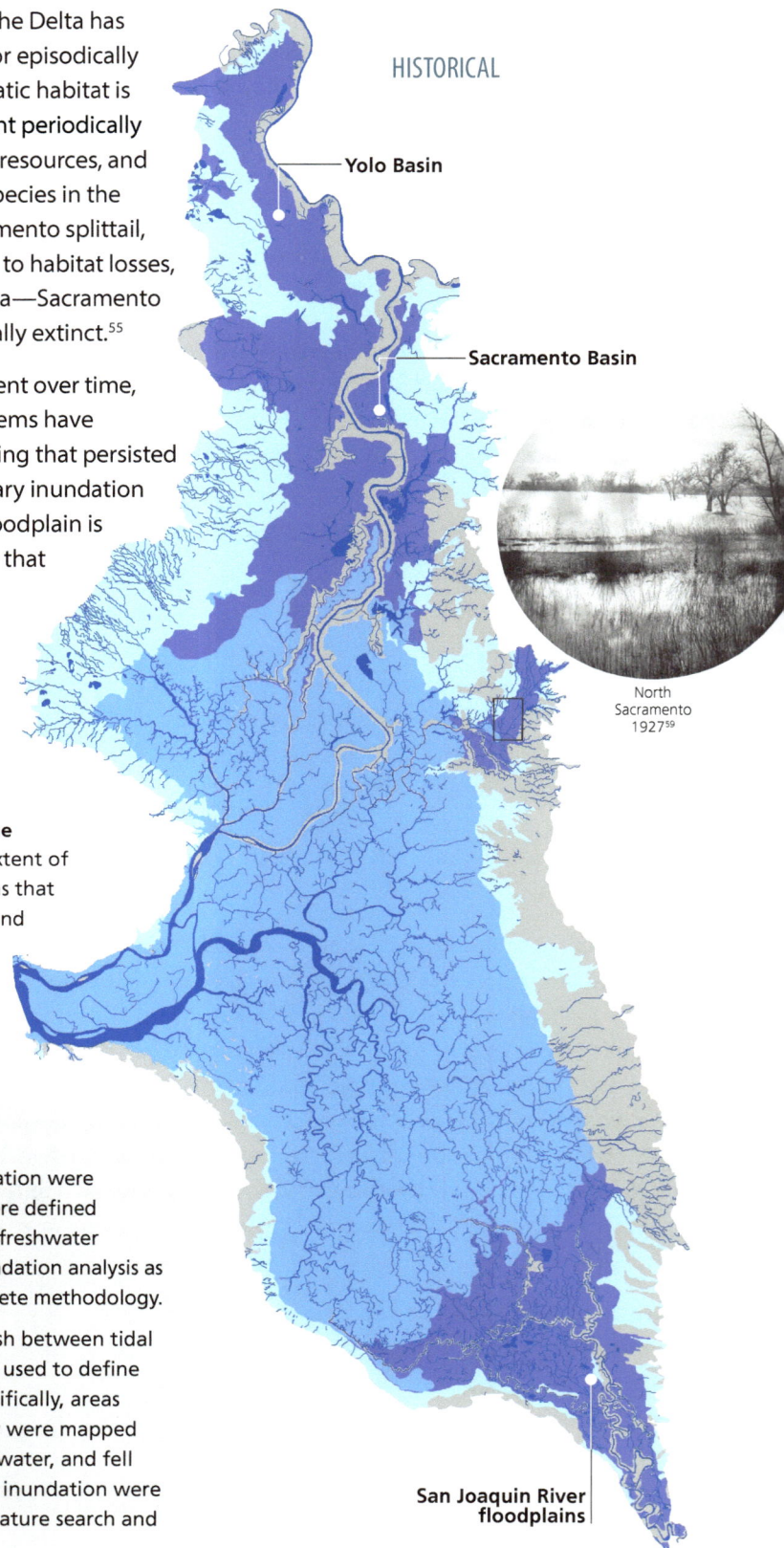

HISTORICAL

Yolo Basin

Sacramento Basin

North Sacramento 1927[59]

San Joaquin River floodplains

MODERN

Yolo Bypass
(1998 flood extent)

SEASONAL SHORT-TERM FLOODING
Short-term fluvial inundation
- intermediate recurrence (~10 events per year)
- low duration (days to weeks per event)
- generally shallower than seasonal long-duration flooding

SEASONAL LONG-DURATION FLOODING
Prolonged inundation from river overflow into flood basins
- low recurrence (~1 event per year)
- high duration (persists up to 6 months)
- generally deeper than seasonal short-term flooding

TIDAL INUNDATION
Diurnal overflow of tidal sloughs into marshes
- high recurrence (twice daily)
- low duration (<6 hrs per event)
- low depth ("wetted" up to 0.5 m)

PONDS, LAKES, CHANNELS, & FLOODED ISLANDS
Perennial open water features (with the exception of historical intermittent ponds and streams)
- recurrence not applicable (generally perennial features)
- high duration (generally perennial features)
- variable depth

salmon reared on Cosumnes River floodplain

salmon reared in Cosumnes River main channel

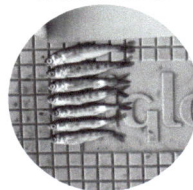

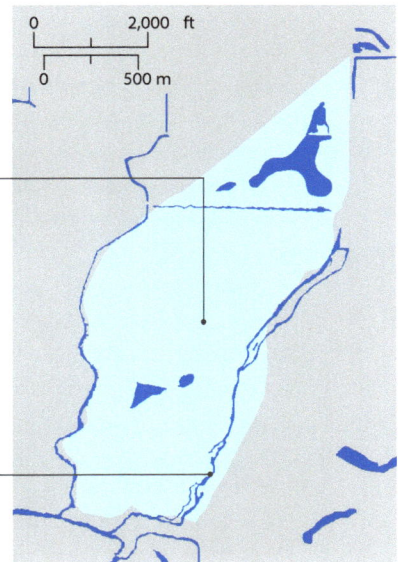

Floodplains support rearing salmon. Juvenile Chinook reared in seasonal floodplain habitats of the Cosumnes River have been found to grow significantly larger than those reared only within the river's main channel.[60] Although seasonally flooded habitat once totalled more than 117,000 ha in the Delta,[61] it is now largely restricted to parts of the Yolo Bypass and Cosumnes River floodplain and totals less than 19,000 ha (a decrease of approximately 85%).

Images of juvenile salmon courtesy Springer Science+Business Media, originally published in Jeffres et al. (2008).

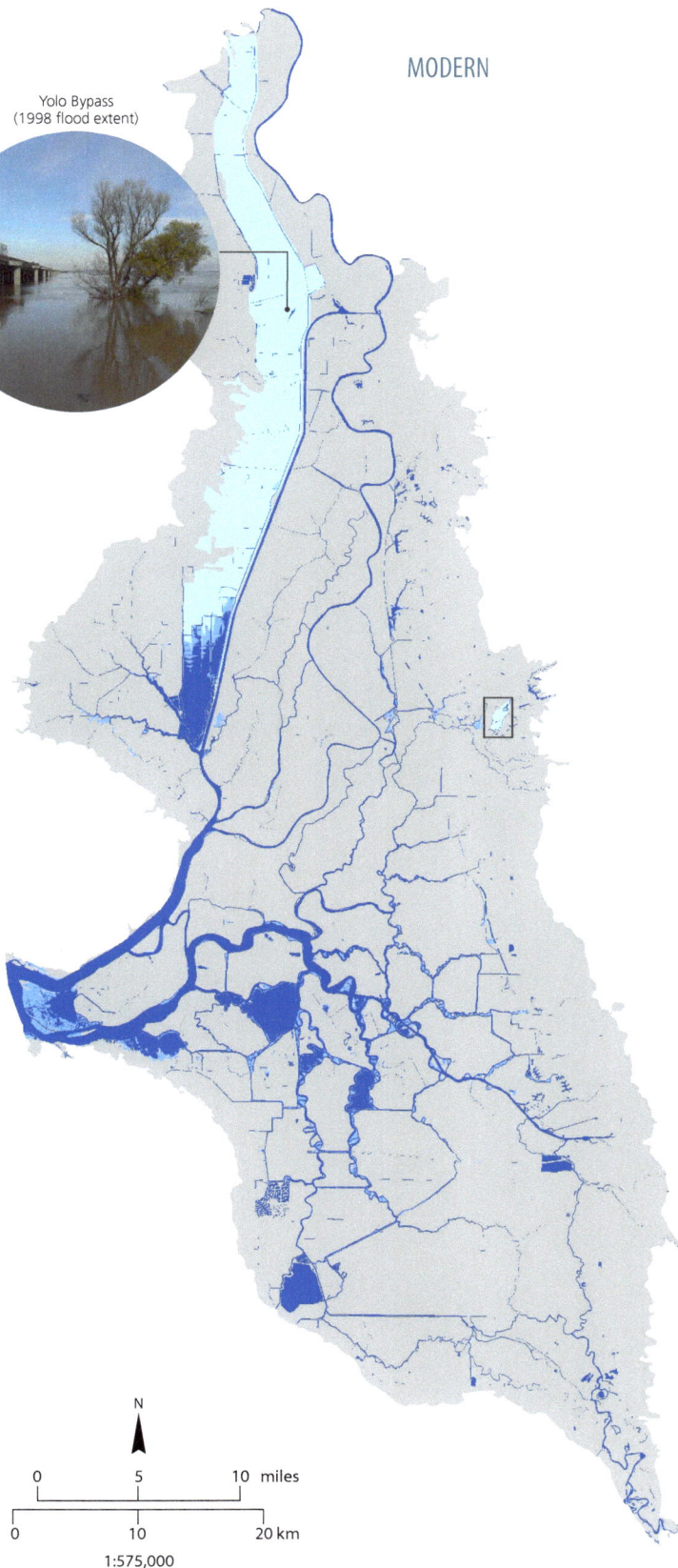

Dramatic loss of seasonally flooded habitats

Native fish are adapted to a complex, variable landscape with extensive aquatic resources throughout the year

The historical Delta exhibited dramatic seasonal variation in flooding (right, top). Seasonal basin flooding in the north Delta, driven by lower-elevation, rain-fed Coast Range streams, tended to occur between December and April. In contrast, elevated flows and flooding in the south Delta were driven by snowmelt, generally began in April, and continued into the summer. This seasonal variation in flooding is reflected in the life histories of the native fish species that evolved here (see bottom of this page and the chart on page 42). Today, a decrease in the extent of inundation across the Delta has been accompanied by a decrease in the spatial-temporal variability of inundation (right).

    SEASONAL SHORT-TERM FLOODING

Short-term fluvial inundation
- intermediate recurrence (~10 events per year)
- low duration (days to weeks per event)
- generally shallower than seasonal long-duration flooding

    SEASONAL LONG-DURATION FLOODING

Prolonged inundation from river overflow into flood basins
- low recurrence (~1 event per year)
- high duration (persists up to 6 month)
- generally deeper than seasonal short-term flooding

    TIDAL INUNDATION

Diurnal overflow of tidal sloughs into marshes
- high recurrence (twice daily)
- low duration (<6 hrs per event)
- low depth ("wetted" up to 0.5 m)

    PONDS, LAKES, CHANNELS, & FLOODED ISLANDS

Perennial open water features (with the exception of historical intermittent ponds and streams)
- recurrence not applicable (generally perennial features)
- high duration (generally perennial features)
- variable depth

Temporal distribution of juvenile Chinook rearing and outmigration (right). The colored bars depict the periods of juvenile rearing and outmigration for each of the four runs of Central Valley Chinook salmon (named for when adults migrate into freshwater; also see page 42). The distinct salmon populations display diverse life-history characteristics that reflect the temporal variability in available habitat across a year **(above)**.[62]

FALL

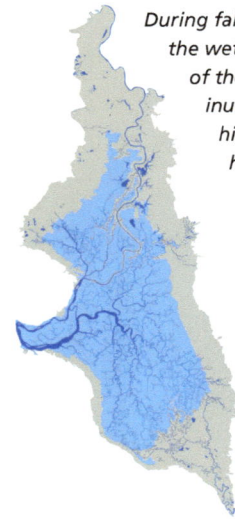

During fall, before the start of the wet season, vast swaths of the historical Delta were inundated by twice daily high tides. The area shown here is the maximum extent historically inundated during spring tides.

HISTORICAL

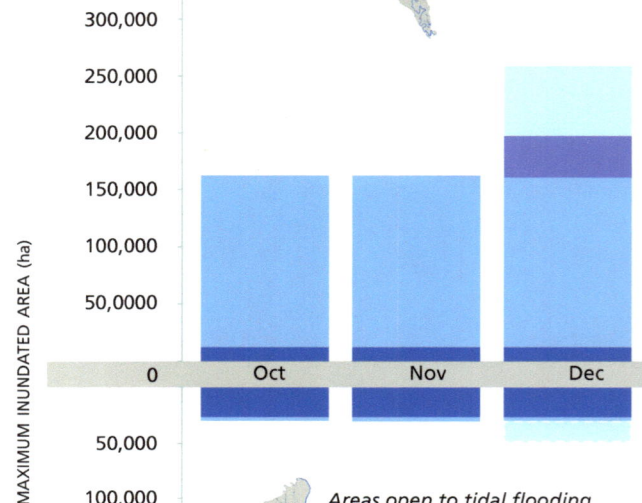

MAXIMUM INUNDATED AREA (ha)

300,000
250,000
200,000
150,000
100,000
50,0000
0 — Oct — Nov — Dec
50,000
100,000
150,000
200,000
250,000
300,000

MODERN

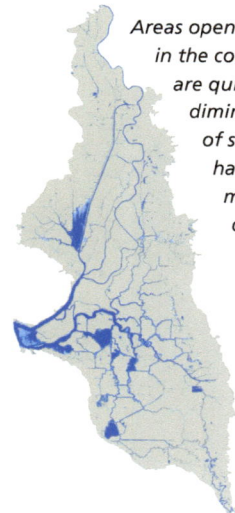

Areas open to tidal flooding in the contemporary Delta are quite limited, greatly diminishing the availability of shallow inundated habitat during dry months, even as deep open water habitats have increased (page 33).

JUVENILE CHINOOK REARING AND OUTMIGRATION

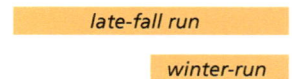

late-fall run

winter-run

WINTER

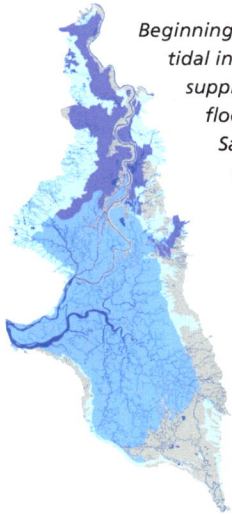

Beginning in the winter, tidal inundation was supplemented with flooding from the Sacramento River, which frequently passed much of its flow into the Sacramento and Yolo Basins. Cache and Putah creeks and also contributed floodwaters.

SPRING

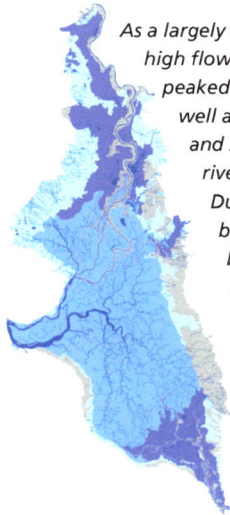

As a largely snowmelt-fed river, high flows on the San Joaquin peaked during the spring, well after winter storms, and spread across the river's floodplain. During wet years, both the north Delta basins and south Delta floodplains would have been inundated during mid-spring, creating a maximum flood extent.

SUMMER

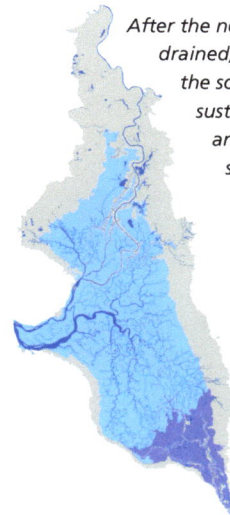

After the north Delta basins drained, inundation in the south Delta was sustained by snowmelt and persisted into the summer, extending the availability of floodplain resources.

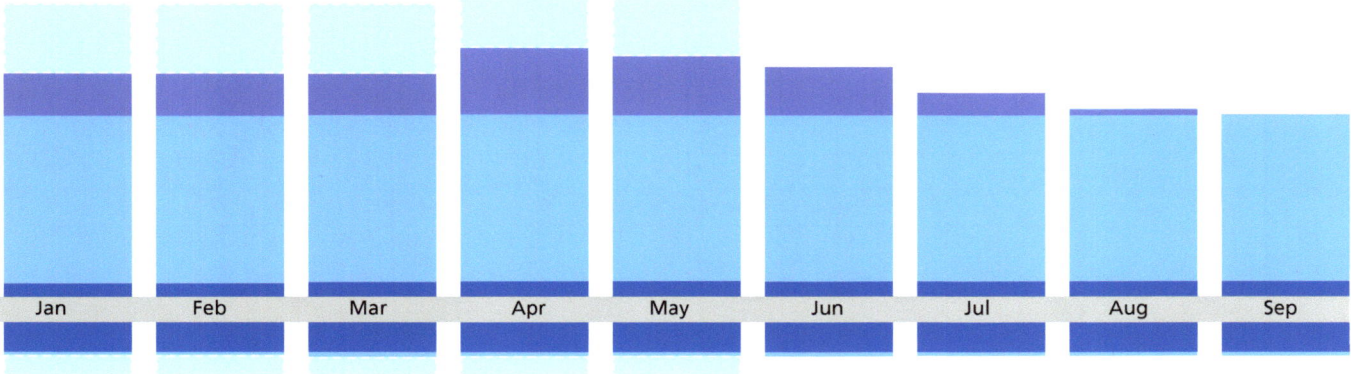

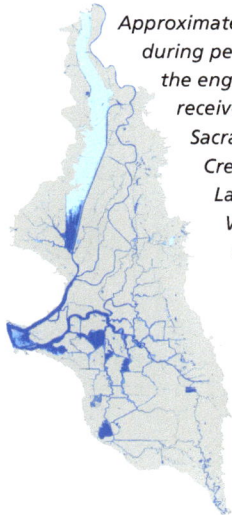

Jan	Feb	Mar	Apr	May	Jun	Jul	Aug	Sep

Approximately every three years, during periods of high flow, the engineered Yolo Bypass receives water from the Sacramento River, Cache Creek, the Knight's Landing Ridge Cut, Willow Slough, and Putah Creek. The floodway is designed to divert this water away from major cities and to quickly deliver it downstream to the Cache Slough Complex.

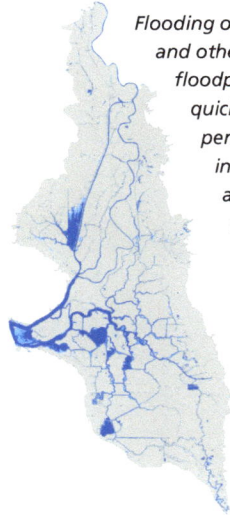

Flooding on the Yolo Bypass and other modern Delta floodplains drains quickly and does not persist for as long into the spring and summer as was common historically.

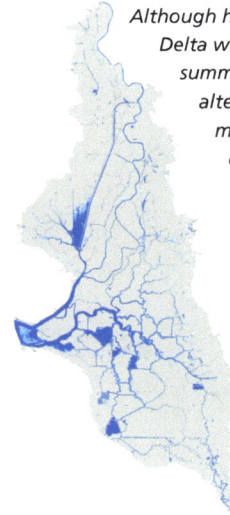

Although historically the South Delta was wettest during early summer, 160 years of flow alterations and channel modifications have changed the timing and magnitude of inundation events in the region. Aquatic habitat in the late spring and summer is now generally limited to small areas of tidal inundation and extensive areas of perennial open water habitats.

fall run

late-fall run

winter-run

spring run

Simple life-history periodicities of BDCP Species of Special Concern (below). Habitat needs vary across fish life-history stages and, therefore, across time. Over the course of a single year, the historical Delta exhibited a great deal of spatial-temporal variability in physical processes/gradients and habitat availability. This variability is reflected in the temporal distributions of fish species that utilize the Delta during one or more phases of their lives. The table reflects modern use of the Delta by fish—it is possible that the historical temporal distributions differed. Migrating adult spring-run Chinook, for example, ascended the San Joaquin River well into the late summer—a pattern that is tied to the availability of snowmelt runoff and sufficient flows from the south Delta to upstream tributaries.[63] There may also once have been (now extinct) summer runs of Chinook and steelhead that migrated in July and August.[64] The table does not include life-history stages that occur predominantly outside of the Delta (like salmonid spawning, which occurs upstream).

■ Life-history stage present in Delta

Species	Life-history stage	O	N	D	J	F	M	A	M	J	J	A	S
Chinook salmon[65]	Fall run adult migration	■	■	■								■	■
	Fall run juvenile rearing and migration				■	■	■	■	■	■			
	Late-fall run adult migration		■	■	■	■							
	Late-fall run juvenile rearing and migration		■	■	■	■	■	■	■	■	■		
	Winter run adult migration			■	■	■	■						
	Winter run juvenile rearing and migration				■	■	■	■	■				
	Spring run adult migration					■	■	■	■	■			
	Spring run juvenile rearing and migration				■	■	■	■	■	■			
Steelhead[66]	Adult migration	■	■	■	■	■	■				■	■	■
	Rearing				■	■	■	■	■				
	Juvenile emigration			■	■	■	■	■	■	■			
Sacramento splittail[67]	Adult upstream migration towards spawning areas		■	■	■	■	■						
	Floodplain/river spawning					■	■	■	■				
	Eggs/embryo and larvae (floodplain/channel margin)						■	■	■				
	Juvenile floodplain use						■	■	■	■			
	Juvenile downstream migration							■	■	■			
Green sturgeon[68]	Juveniles (Delta/Bay)	■	■	■	■	■	■	■	■	■	■	■	■
	Spawning migration (Bay/Delta)					■	■	■	■	■			
	Post-spawn adults (River/Delta)						■		■	■	■		
	Mature adults (Ocean/Delta)	■	■	■	■	■	■						■
White sturgeon[69]	Juveniles	■	■	■	■	■	■	■	■	■	■	■	■
	Spawning migration					■	■	■	■	■			
Pacific lamprey[70]	Adult migration	■	■	■	■	■	■	■					
	Ammocoetes (larval lamprey)	■	■	■	■	■	■	■	■	■	■	■	■
	Metamorphosis to juveniles									■	■	■	■
	Juvenile outmigration	■	■	■								■	■
River lamprey[71]	Adult upmigration	■	■	■	■	■	■	■					
	Juvenile outmigration (congregation in Delta)								■	■	■		
Delta smelt[72]	Egg/embryo (sandy-gravel channel edge)					■	■	■	■				
	Yolk-sac/First-feeding larvae (offshore tidal freswater)					■	■	■	■				
	Fin-fold larvae (offshore tidal freswater)						■	■	■				
	Metamorphosing larvae (offshore tidal freshwater & LSZ)							■	■	■			
	Juveniles (offshore tidal freshwater & LSZ)	■	■						■	■	■	■	■
	Migrating adults (offshore tidal freshwater)			■	■	■	■						
	Spawning (tidal freshwater)					■	■	■	■				
Longfin smelt[73]	Spawning		■	■	■	■	■						
	Eggs		■	■	■	■	■						
	Larvae			■	■	■	■	■					
	Juveniles (primarily in San Francisco and Suisun bays)	■	■	■	■	■	■	■	■	■	■	■	■

Chinook salmon in the lower American River **(top)** and at a fish hatchery **(bottom)**.

Photo Credits: Dan Cox, USFWS; Steve Martarano, USFWS

Marshes directly influenced the character and quality of aquatic habitats

There has been a 73-fold reversal in the ratio of marsh to open water area in the Delta

The Delta has shifted from a system of tidal channels surrounded by marsh to one dominated by leveed open water with little marsh influence. Aquatic habitat in the historical Delta was strongly linked to wetlands, which contributed to productivity and turbidity, and influenced the hydrology, structure, and chemistry of adjacent aquatic habitats. Some fish species likely accessed the marsh plain and marsh edge directly, while others may have benefited from the export of nutrients, food, and organic matter from marshes. The extent to which marshes benefitted native species is hard to determine because so little marsh remains today.

Marsh and open water habitat adjacencies in the historical (right) and modern (far right) Delta. The marsh-open water edge is color-coded by the size of the adjacent marsh. Both the ratio of marsh to open water and the total length of marsh-open water edge have decreased dramatically. These figures and tables do not include an estimated additional ~3,800 km of historical marsh-water edge associated with the smallest, unmapped channels.

Methods: Marsh to open water ratio and edge

For the analyses on this page, we isolated all areas mapped as open water and marsh, regardless of their tidal status, connectivity, or form. Since habitat type maps represent average dry-season conditions, seasonally and tidally inundated areas are not included within the area mapped as open water. Linear areas where the two habitat types were mapped as adjacent to one another are identified as the open water-marsh edge. This edge was then classified by the size of the contiguous area of marsh from which it was drawn.

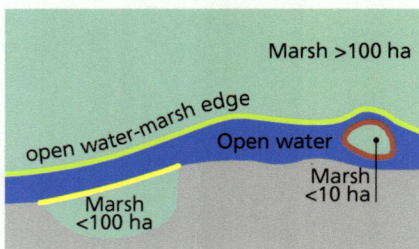

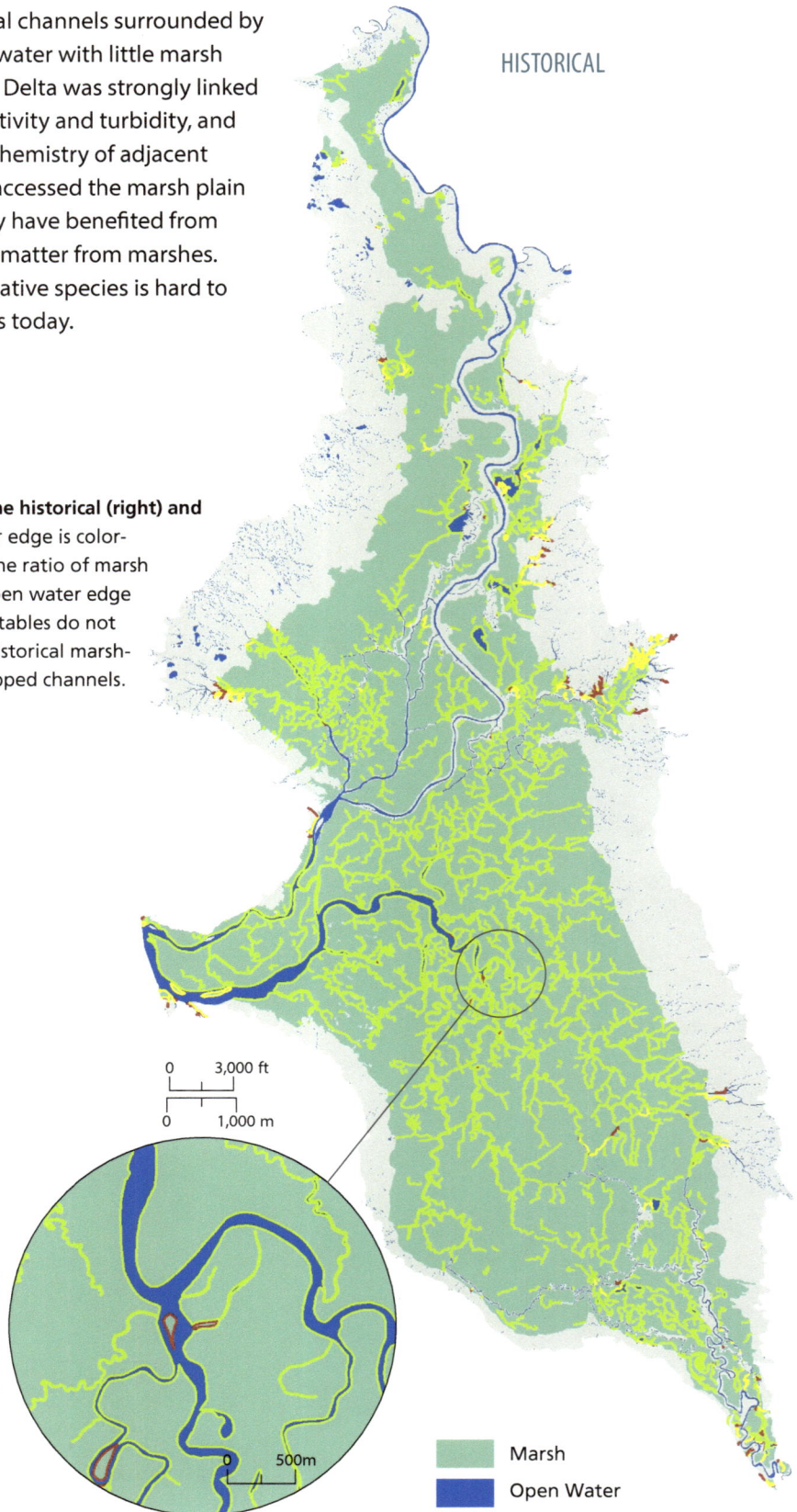

HISTORICAL

0   3,000 ft

0   1,000 m

Marsh >100 ha

open water-marsh edge

Open water

Marsh <100 ha

Marsh <10 ha

0   500m

Marsh

Open Water

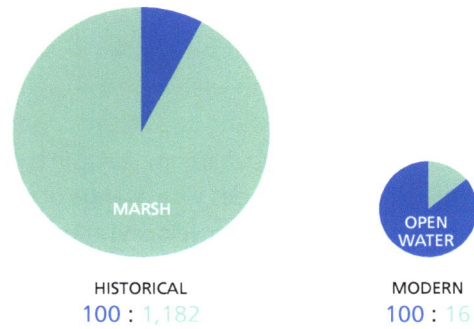

MODERN

Liberty
Island

Sherman
Island

| 0   3,000 ft |
| 0   1,000 m |

N

| 0 | 5 | 10 miles |
| 0 | 10 | 20 km |

1:575,000

Habitat type	Total area (ha)	
	Historical	*Modern*
Marsh	193,224	4,296
Open water	16,344	26,554

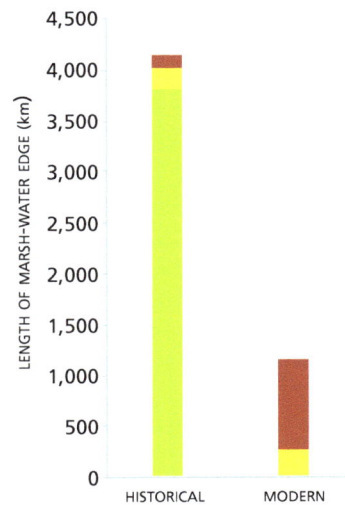

MARSH

OPEN
WATER

| HISTORICAL | MODERN |
| 100 : 1,182 | 100 : 16 |

The reversal in marsh to open water area ratio over time (above) is the result of a 98.7% decrease in the area of marsh and a 62.5% increase in the area of open water. Where historically the Delta was characterized by narrow channels embedded within large areas of marsh, today we find tiny marshes embedded within large areas of open water.

Marsh-water edge marsh size class (ha)	Marsh-water edge length (km)	
	Historical	*Modern*
>100 ha	3,823	31
10 - 100 ha	202	236
0 - 10 ha	112	874
TOTAL	**4,137**	**1,142**

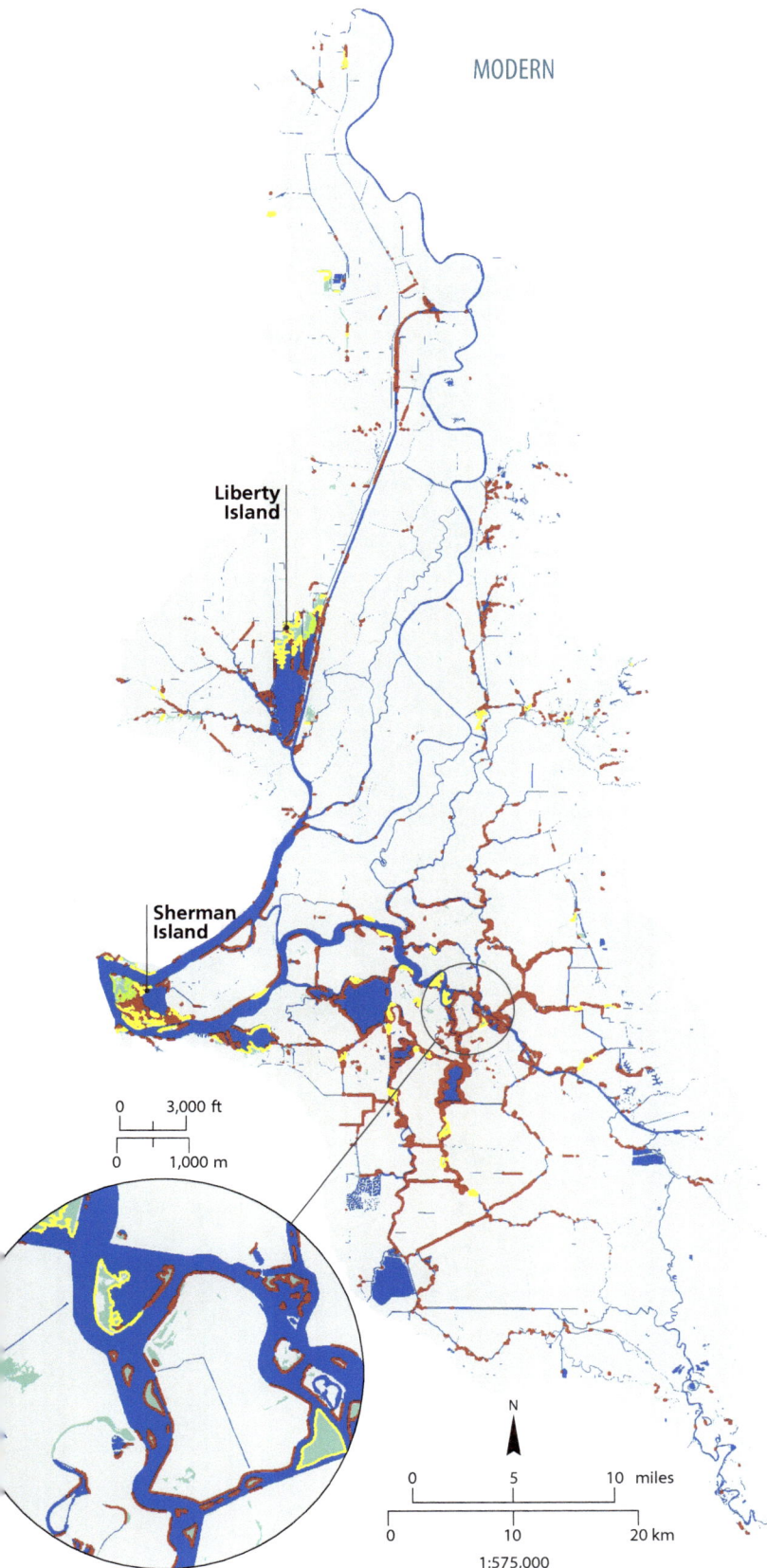

LENGTH OF MARSH-WATER EDGE (km)

HISTORICAL   MODERN

Despite fragmentation (which increases marsh edge length), the length of marsh-water edge has decreased by more than 72%. Historically there was over 3500 km of interface between open water and large (>100 ha) marsh patches The present day edge is largely associated with marsh patches <10 ha in size.

Complex dendritic channel networks likely provided high productivity habitat for fish

Most dendritic channels are now gone, especially in the central Delta

As Delta marshes were diked, connections were severed to the channel networks that wove through them. These dendritic lower-order tidal channels (also known as "dead-end" or "blind channels") that terminated within the wetland were once the capillary exchange system between the wetland and aquatic areas, promoting both food web productivity and spatial complexity in habitat conditions. They provided native fish species with a range of gradients (e.g., temperature, turbidity, and water velocity) at both large and small scales. Dendritic channel networks offered channel complexity and higher turbidities, which provided refuge for certain species. Channels that branched through the marsh may have been particularly important for salmonids because they provided access to and export of invertebrates from the marsh plain,[68] physical cover and turbidity for refuge, and slow moving water for energetic refugia. The larger, looped channels that characterize the Delta today allow water to move through and mix more quickly, with less diversity in residence time and less heterogeneity in channel habitat. The lack of large wetlands connected to channels means that there is little exchange of organic matter, organisms, or sediment between these ecosystems.

Comparing the historical (right) and modern (far right) landscape. While the skeletal framework of looped mainstem channels remains largely similar (red), the branching networks of dendritic channels (green and yellow) are mostly gone.

Methods: Classifying channel types

Channel reaches were manually classified using the following definitions:

Dendritic: tidal channel reaches connected to the tidal source by only one non-overlapping path

Looped: tidal channel reaches connected to the tidal source (the Delta mouth) by two independent and non-overlapping paths

Fluvial: channel reaches connected to the tidal source, but upstream of the approximate limit of bidirectional tidal flows (during spring tides in times of low river stages) AND tidal reaches between upstream perennial fluvial reaches and downstream looped reaches

Detached: channel reaches without a direct connection to the tidal source (through the larger channel network)

Dendritic channels (segmented at 100 m intervals) were classified into those adjacent to marsh and those non-adjacent to marsh, based on the habitat-type polygon closest to the channel centerline.

HISTORICAL

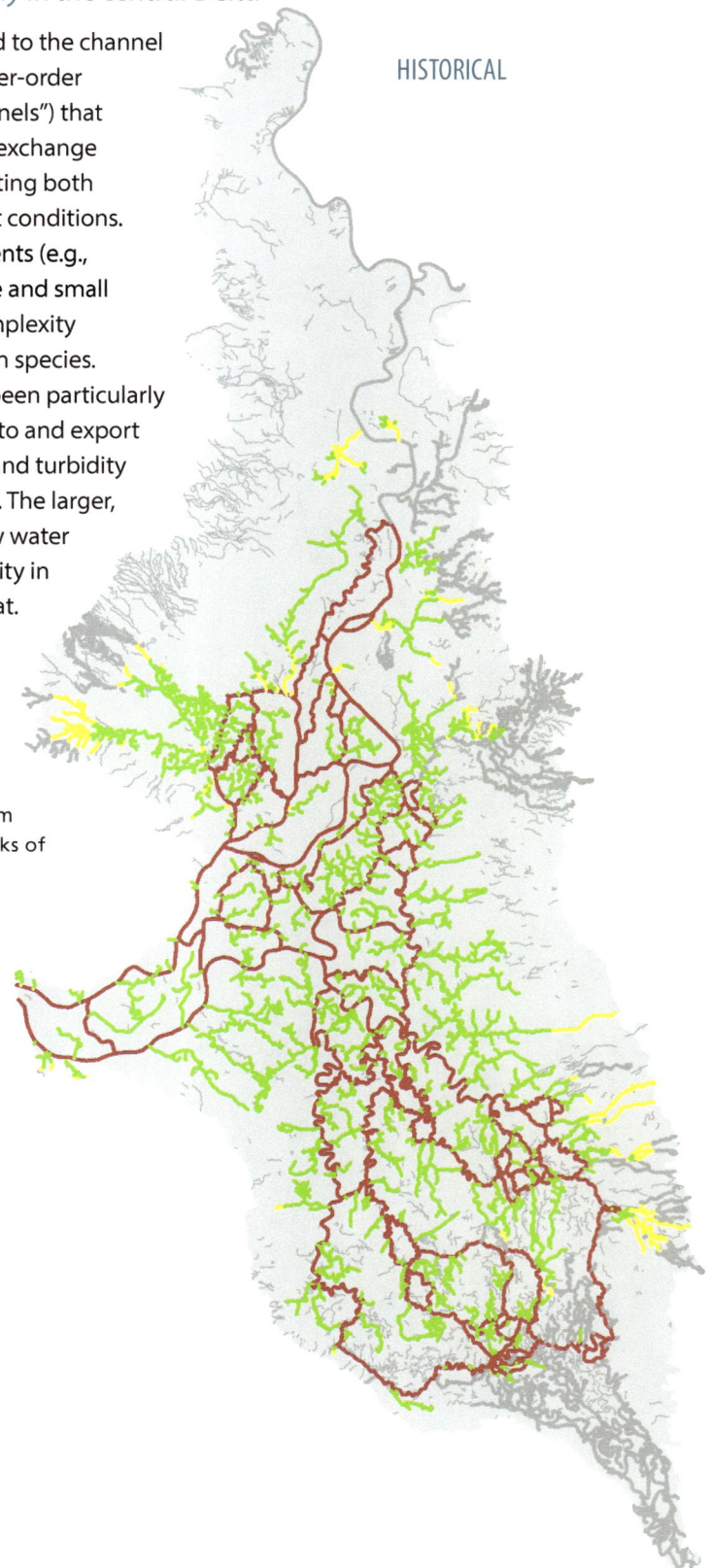

MODERN

Liberty
Island

Sherman
Island

N

| 0 | 5 | 10 miles |

| 0 | 10 | 20 km |

1:575,000

Channel classification		Channel length (km)	
		Historical	*Modern*
〰	Dendritic channels adjacent to marsh	1,151	84
〰	Dendritic channels not adjacent to marsh	153	255
〰	Looped Channels	754	768
〰	Fluvial		
〰	Detached	2,225	298
	TOTAL	4,283	1,404

Most channels in the Delta today are looped. The length of this kind of channel has slightly increased (due to channel cuts), while the length of dendritic tidal channels has decreased by more than 74%. Where dendritic channels do exist, they are generally not part of marshes—the length of dendritic channels adjacent to marsh has decreased by 93%. These figures and tables do not show or account for the approximately 1,900 km of estimated unmapped, low-order dendritic channels in the historical Delta.

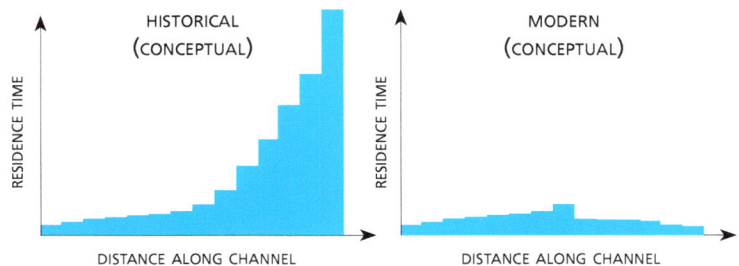

HISTORICAL
(CONCEPTUAL)

RESIDENCE TIME

DISTANCE ALONG CHANNEL

MODERN
(CONCEPTUAL)

RESIDENCE TIME

DISTANCE ALONG CHANNEL

DENDRITIC CHANNEL SYSTEM

CHANNEL CUT

Historically, the complex structure of Delta channels established gradients in residence time, a pattern heavily altered in the modern Delta (after Chris Enright, Delta Science Program). Historically, small low-order tidal creeks had high residence times, which allowed phytoplankton to accumulate and created net autotrophic conditions. Deeper sloughs, by contrast, had shorter residence times which created net heterotrophic conditions. The increased connectivity of modern channels in the Delta has led to homogenization of residence time across channel networks, increasing the reach of tidal excursion within channel networks and decreasing the occurrence of small channels with high residence time. The relationship between residence time and primary productivity in the modern Delta has been additionally complicated by the introduction of the overbite clam.

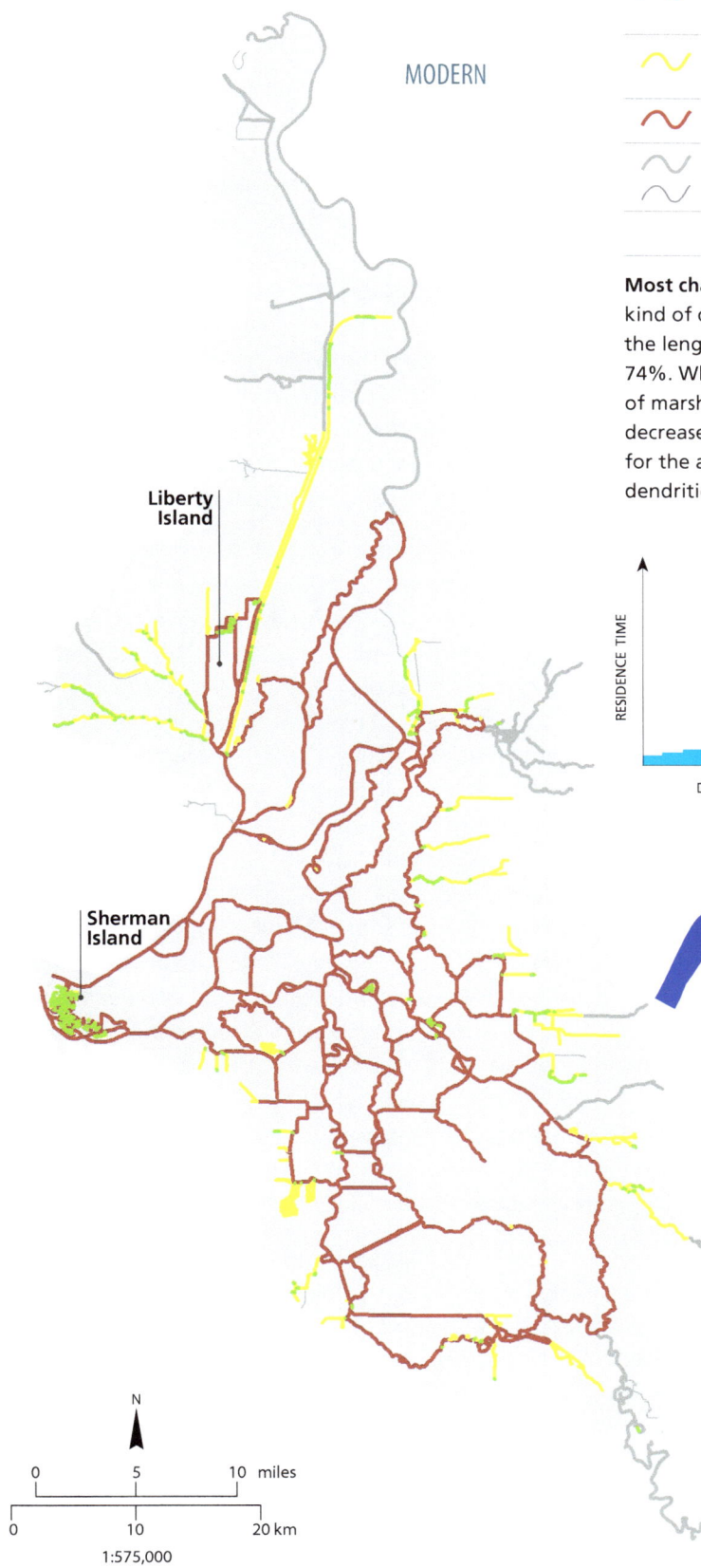

5. Life-History Support for Marsh Wildlife

Freshwater marshes dominated the Delta landscape historically. Enormous expanses of regularly inundated, highly connected, productive, and structurally complex marsh sustained large wildlife populations. Although much of the marsh was dominated by tules, overall the marsh supported a rich assemblage of both perennial and annual plant species that added to the marsh structure and complexity.[1] In this report, we use the word "marsh" to describe both tidal and non-tidal freshwater emergent wetlands, which can include non-herbaceous species, such as willows. Diverse plant species produced large quantities of seeds that accumulated as extensive seed banks in the sediment.[2] In tidal freshwater marshes both larval and adult insects were key primary consumers.[3] The amount of plant production directly influenced the potential to support secondary consumer populations by providing organic matter for detritivores, contributing to habitat structure, and other mechanisms.[4] The abundant food resources of Delta marshes supported many wetland and terrestrial vertebrates. Some terrestrial and semi-terrestrial species were restricted to the freshwater marsh, while others used it as one of several habitat options, as a migration corridor, or for a part of their life history (such as for dry-season foraging).[5]

A diverse and dynamic community of native wildlife, including humans, flourished within the marshes of the historical Delta. This community included resident birds and mammals such as rails, herons, bitterns, songbirds, mice, shrews, and voles.[6] Tidal freshwater marshes are thought to support the largest and most diverse populations of birds of any wetland type.[7] Waterbirds such as coots, moorhens, grebes, ducks, geese, and swans inhabited the channels and ponds within the Delta marshes, taking advantage of the food and shelter that marsh proximity provided. Some waterbirds also used the marsh to forage, rest, or breed (see Chapter 6). The Delta supported abundant beavers, river otters, and mink and was a major population center for these species.[8] The shallow ponds, blind channels, and backwaters of the marsh provided slow-moving habitat for littoral fish such as tule perch and the now extinct thicktail chub. Some fish inhabited the smaller marsh channels and may have ventured further into the marsh as flooding conditions allowed (see Chapter 4).[9] Tree frogs, pond turtles, California red-legged frogs, and giant garter snakes that used the marsh were likely limited to areas close to upland and seasonal wetland habitats.[10] In addition many terrestrial species, notably tule elk, but also antelope, deer, coyotes, and bears, used the marsh opportunistically to supplement foraging or escape predation and extreme conditions (see Chapter 8).[11] Raptors, including Northern Harriers and White-tailed Kites, hunted in the marsh. Compared to high salinity tidal marshes, freshwater marshes are thought to have high wildlife diversity, but low endemism.[12] However the Delta did support several endemic plants and a few regionally restricted vertebrates including the giant garter snake and Modesto Song Sparrow.[13] Finally, indigenous people benefited from and managed for this wildlife diversity, relying on the extensive marshes for food and materials.[14]

The considerable heterogeneity expressed by Delta marshes provided structural complexity and niche diversity to support different species. Gradients in physical characteristics such as tidal energy, river flow, and salinity, as well as subtle local variations in topography and microclimate, provided a variety of habitat features that supported wildlife under different conditions (e.g., seasonal cycles, floods, drought, temperature extremes, turbidity). These gradients also supported different species in different places, and fostered genotypic and phenotypic diversity within species. The character of the Delta marsh was particularly variable along its latitudinal gradient. Largely due to its distance from the mouth of the Delta and to

Marsh wildlife (right). Top to bottom: North American river otter, Delta tule pea, Virginia Rail, giant garter snake.

Photo Credits: Unknown, USFWS; Mark Fogiel, CalPhotos; Tom Talbott, Creative Commons; Dave Feliz, Wikipedia Commons

riverine influences, the north Delta flood basins contained broad zones of both tidal and non-tidal freshwater marsh that were relatively free of channels and supported dense stands of tules over ten feet tall. Channel density and sinuosity in the central Delta was greater than in the less tidally dominated northern and southern parts of the Delta because of the gradation in tidal prism. Willows were a significant component of the western-central Delta marshes, which were characterized by willow-fern-tule associations. The marshes of the south Delta were a mosaic of small ponds, patches of tule, willow thickets, rushes, grasses, and sedges[15] dependent on fluvial geomorphic influences from the San Joaquin River and tributaries. In addition to gradients, disturbances (including flood, drought, animal damage, and fires) maintained heterogeneity within the marsh. By knocking back vegetation, these disturbance mechanisms allowed disturbance-tolerant plants to grow and created small open water habitats (duck puddles) that supported waterfowl and littoral fish. The north Delta in particular supported many such small ponds.

The staggering loss of marsh in the Delta, combined with changes in connectivity and habitat quality, has led to tremendous loss of wildlife support. Over 97% of the historical marsh is now gone. What little marsh remains consists primarily of small patches surrounded by deep channels, artificial levees, and agriculture. Much of the marsh in the modern Delta is the result of accidental restoration via levee failure, and is relatively young in age (decades old rather than centuries). Marshes in the Delta no longer span broad, continuous gradients; instead, isolated patches occupy narrow spots along these gradients. Many modern marsh patches are small islands—often the cut-off tips of once larger marshes—now surrounded by riprapped levees and deep channels. The size and isolation of existing marsh patches severely limits the wildlife populations the marsh can support. The Delta's waters no longer inundate surrounding wetlands, limiting exchange of nutrients, organic matter, and dry-season freshwater input.

Fragmented wetlands support smaller wildlife populations because of increased edge effects, with reduced population viability and greater probability of extirpation within habitat fragments.[16] With few patches large enough to support self-sustaining populations, marsh wildlife in the Delta is particularly vulnerable to catastrophic events. The complex channel networks that were associated with these marshes historically cannot be adequately expressed in the small remaining habitat patches. In addition to these effects of fragmentation, the habitat quality of the remaining marsh patches has been altered by non-native invasive species, which compete with and prey upon native species, and by changes in water quality due to agricultural and urban runoff and habitat alteration.[17] Species that relied on the marsh historically, including waterfowl, giant garter snakes, and Tricolored Blackbirds, now increasingly rely on other habitats, including agricultural fields and blackberry thickets. Managed wetlands are critical to wildlife support in the Delta, providing habitat for wintering and nesting waterfowl (see Chapter 6) but they do not support the full native marsh wildlife community, and are often not hydrologically connected to the larger Delta system.

The freshwater marshes of the Delta were unique and extraordinarily valuable to wildlife. As part of an interior inverted delta in a Mediterranean climate, unparalleled in size within the state, the freshwater marsh in the Delta offered unique benefits to wildlife. Because so little of this habitat remains intact it is difficult to comprehend what has been lost. The majority of the Delta historically supported native marsh wildlife; now few places in the Delta do.

The few marshes left in the Delta are small

The current average patch size is several hundred times smaller than in the historical Delta

The area of marsh in the Delta has been reduced by 97%. The size of marsh patches today is small relative to the scale at which important physical and ecological processes occur. Large marsh patches are more likely than small patches to have well-developed channel systems and a range of physical and ecological features. Large patches spanned considerable heterogeneity in inundation patterns, vegetative structure, and geomorphic setting. Here we look at spatial patterns of marsh patches based on parameters relevant to marsh wildlife, such as intertidal rails (for more details see Appendix A, pages 89-90). Spautz and Nur (2002) observed that Black Rails were more consistently detected in marshes greater than 100 ha. Only three marsh patches larger than 100 ha remain in the Delta (compared with 14 historically).

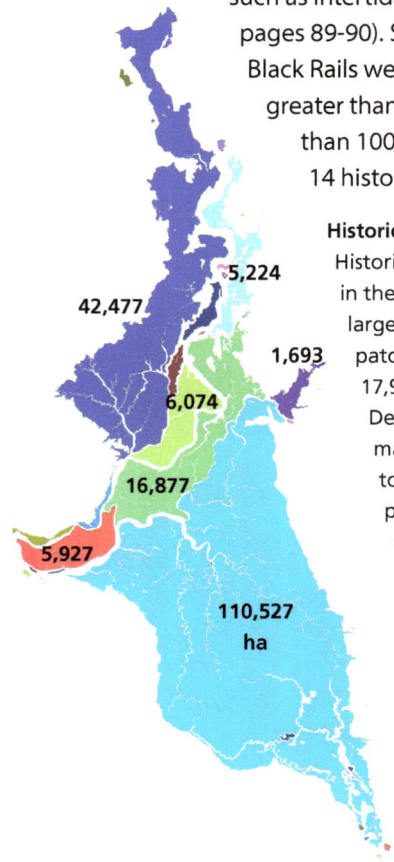

HISTORICAL

Historical marsh patches with patch sizes (left).
Historically, there were 43 marsh patches in the Delta (14 of which were larger than 100 ha), with a mean patch size of 4,494 ha (SD = 17,956 ha). In the modern Delta there are many more marsh patches (1,211 in total), but they have a mean patch size of only 4 ha (SD = 24 ha), and only three are larger than 100 ha. Average patch size today is thus several hundred times smaller than it was historically. The largest single historical patch (110,527 ha) spanned the entire south and much of the central Delta.

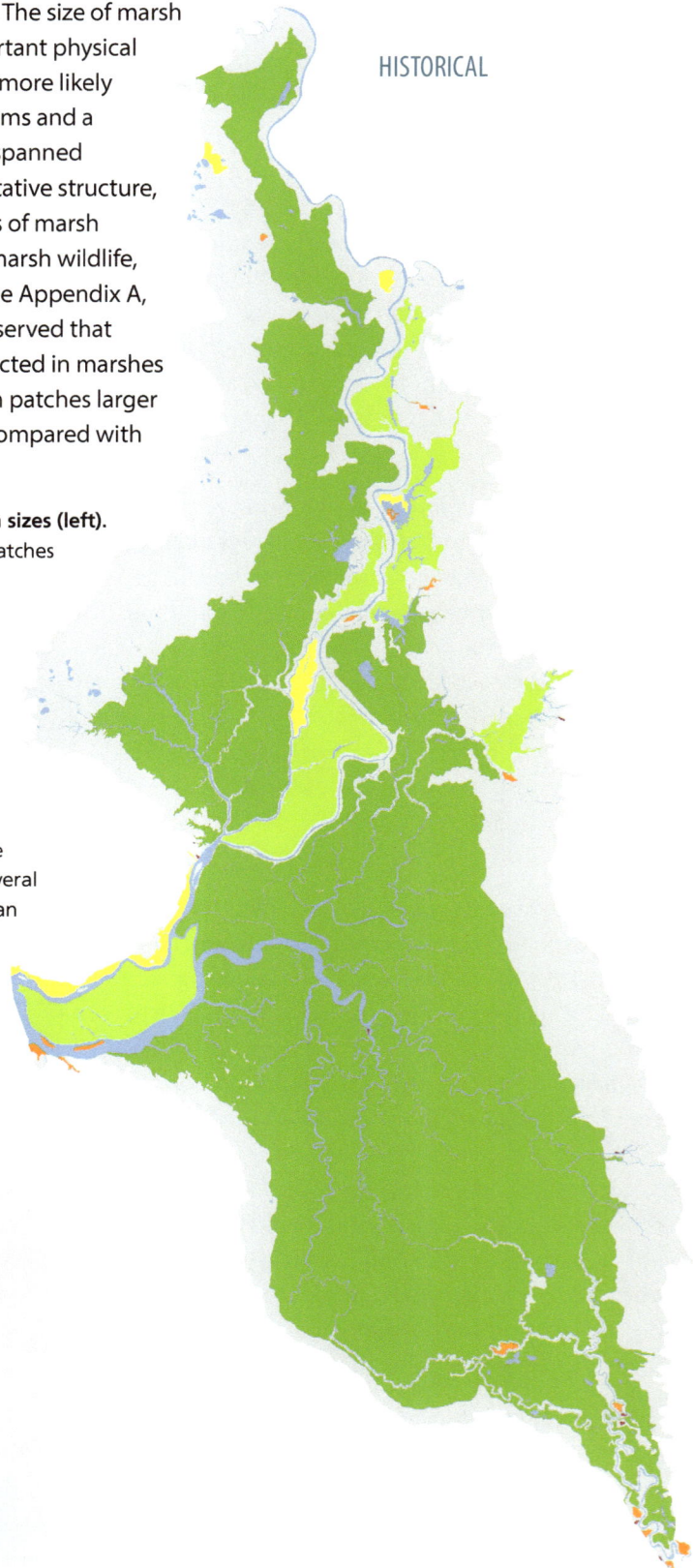

5,224

42,477

1,693

6,074

16,877

5,927

110,527 ha

Methods: Patch size

individual polygons are grouped into one patch if less than 60 m apart

<60 m

polygons greater than 60 m away from each other are considered to be separate patches

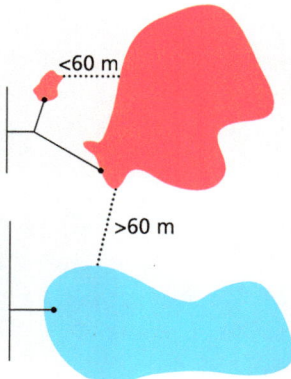

>60 m

MODERN

Marsh patch size class (ha)		Total area (ha)	
		Historical	*Modern*
🟥	<= 10 ha	46	1,427
🟧	10 - 100	702	1,757
🟨	100 - 1,000	2,489	1,154
🟩	1,000 - 10,000	20,105	0
🟢	>10,000	169,881	0
	TOTAL	**193,224**	**4,338**

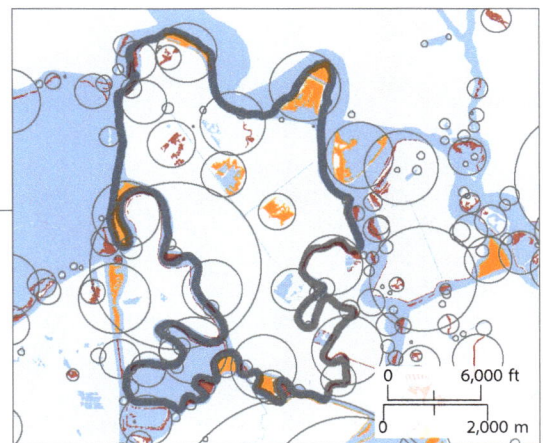

HISTORICAL   MODERN

Historical marsh was composed of large patches, unlike marsh in the modern Delta **(above)**. Even today's largest patches—Liberty Island and Sherman Island—are not very big by historical standards **(left)**. The total extent of marsh is <3% of the historical area.

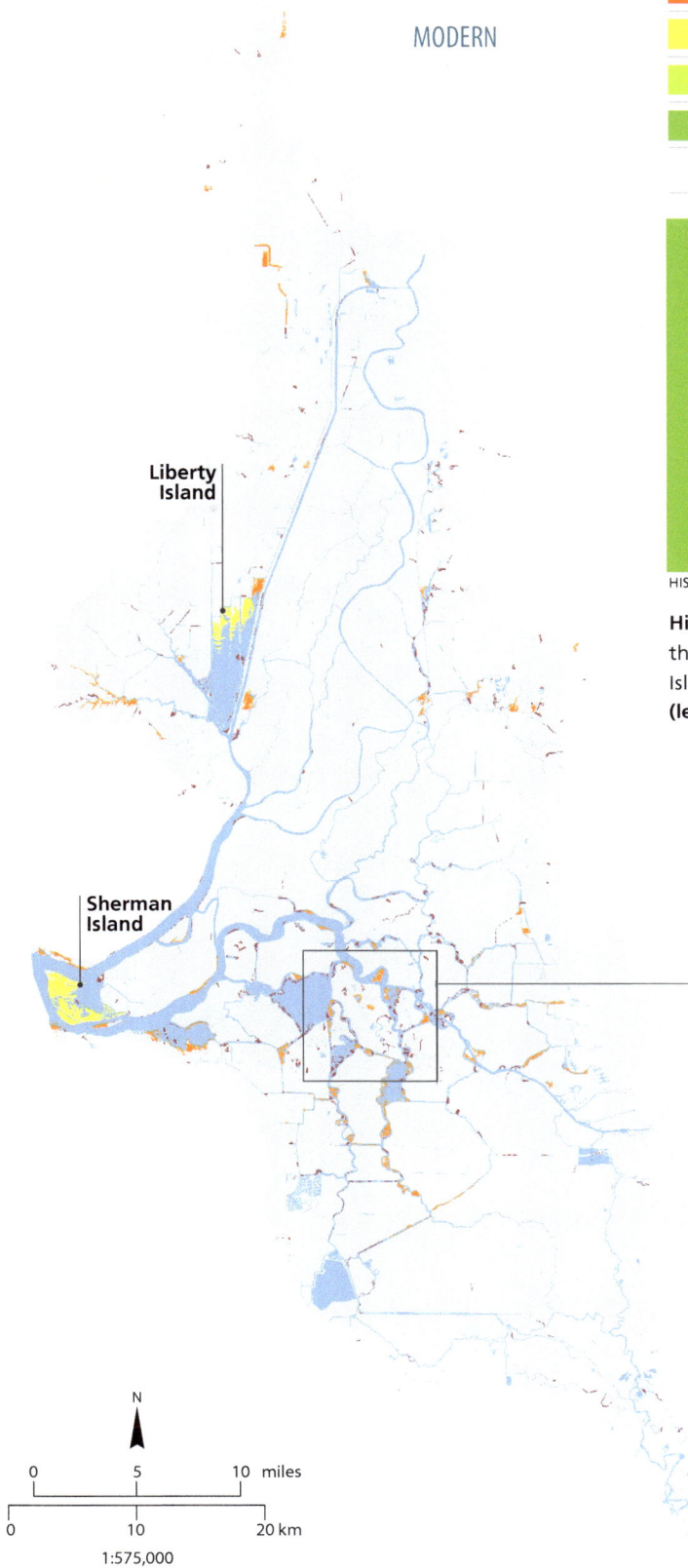

Today there are hundreds of tiny marsh patches scattered throughout the Delta (above), each represented here with a circle indicating the extent of each individual patch. Most of these are fringing marshes along channels or the tips of former large marsh islands. These scattered small remnants were cut off by the excavation of levees and widening of channels over a century ago. For example, the area outlined in grey above shows what was once a continuous historical marsh island that is now fragmented into small pieces.

Existing marshes are isolated

Average distance from a marsh patch to the nearest large marsh has increased more than 50-fold

Continuous marsh habitat is essential for dispersal, foraging, gene flow, and resilience to disturbance for marsh wildlife populations. Marsh patches in the modern Delta are now isolated from one another, fragmenting populations of marsh wildlife. Historically, all marsh patches were within 1.62 km of a large (>100 ha) marsh, with the average distance to a large patch being 0.29 km (SD = 0.40 km). In the modern Delta, the average distance to a large patch is 19.3 km (SD = 11.08 km)—two orders of magnitude farther—with a maximum distance of 61.4 km. Wildlife in small, isolated patches are less likely to disperse successfully. Populations that are lost from these patches due to catastrophic or stochastic events are less likely to be re-established due to low re-colonization rates.[18] In the long run, isolated and small populations can lose genetic diversity.

HISTORICAL

Patterns within the historical (right) and modern landscape (far right). Historically, marsh patches were close together (generally within 1 km). This landscape configuration allowed wildlife movement to maintain diversity within these small patches. Large marsh patches were separated from one another by wide stretches of river or associated riparian forest. In the modern Delta, marsh patches are significantly smaller, and more isolated.

Methods: Nearest large neighbor distance
Measuring each patch's distance to another patch of at least 100 ha

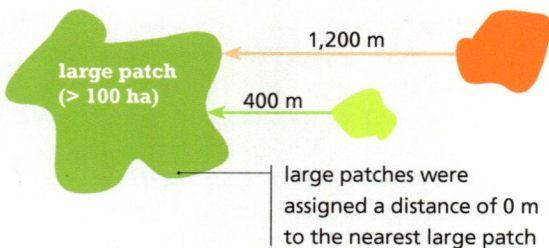

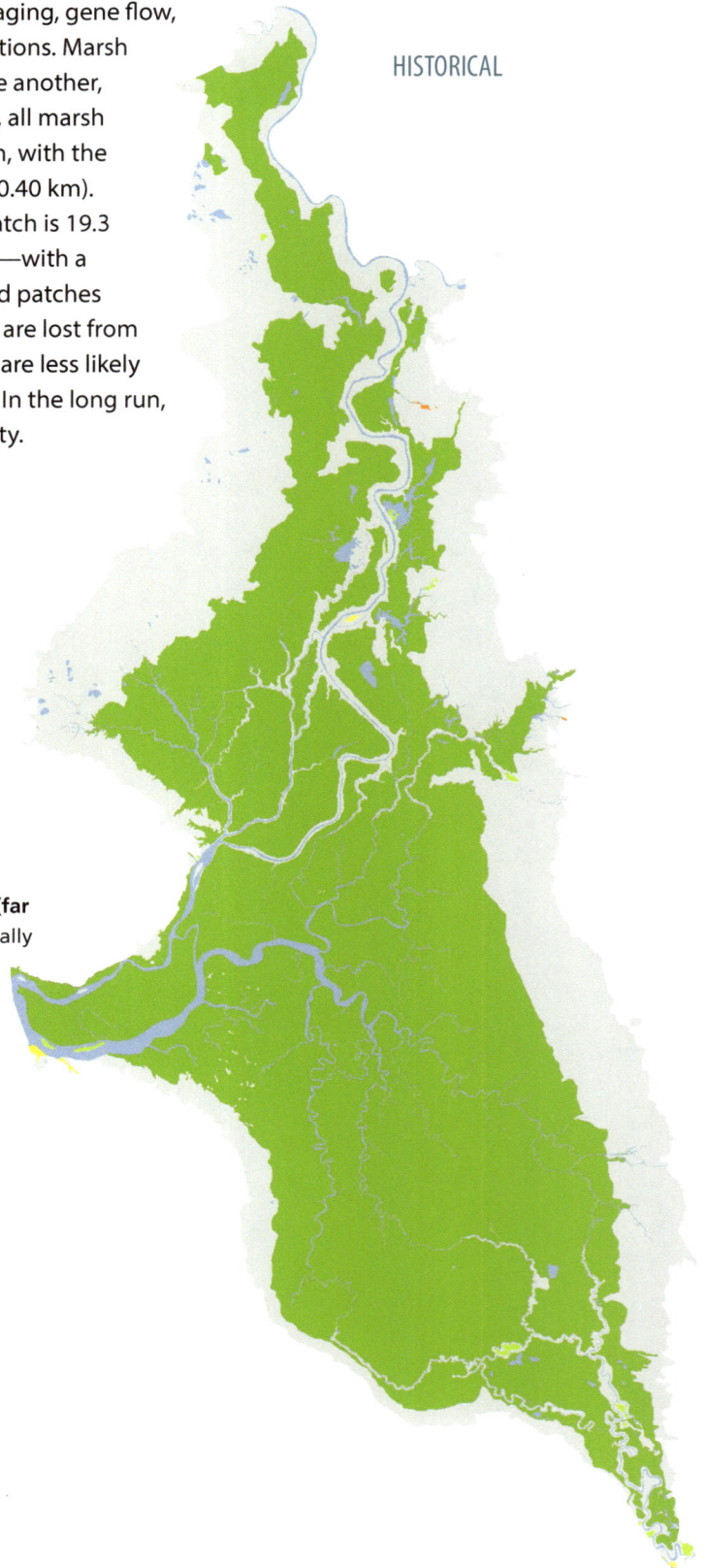

1,200 m

large patch (> 100 ha)

400 m

large patches were assigned a distance of 0 m to the nearest large patch

MODERN

Distance from nearest large neighbor (m)		Total area (ha)	
		Historical	*Modern*
	<= 100 m	192,523	1,161
	100 - 500	500	143
	500 - 1,000	169	87
	1,000 - 10,000	32	630
	>10,000	0	2,317
	TOTAL	**193,224**	**4,338**

HISTORICAL   MODERN

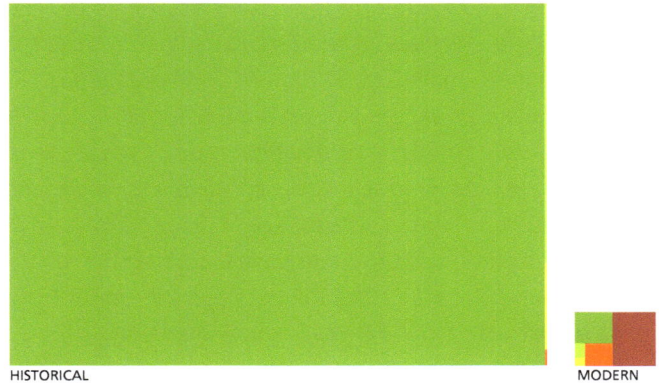

Most modern marsh patches are isolated from significantly sized neighbors **(above)**. Creation of larger marsh patches in the Delta would increase habitat value for the surrounding marshes.

Loss of connectivity. There are few marsh patches between Liberty Island and Sherman Island (today's largest patches) that might serve as stepping stones for movement of wildlife between these two areas **(left)**. In the Central Delta, the patches that do exist are not only small, but are also isolated from the large patches **(below)**. Extensive areas of the south and north Delta are largely devoid of marsh.

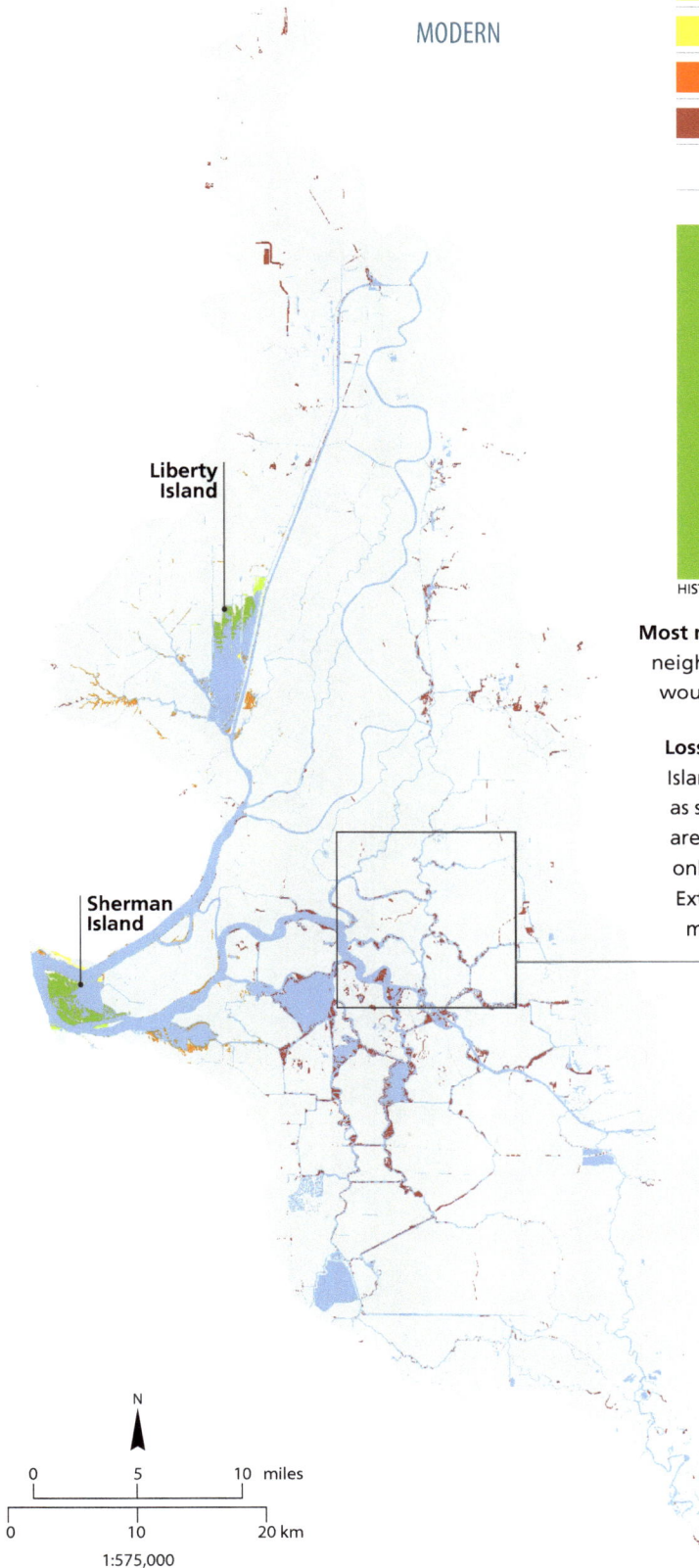

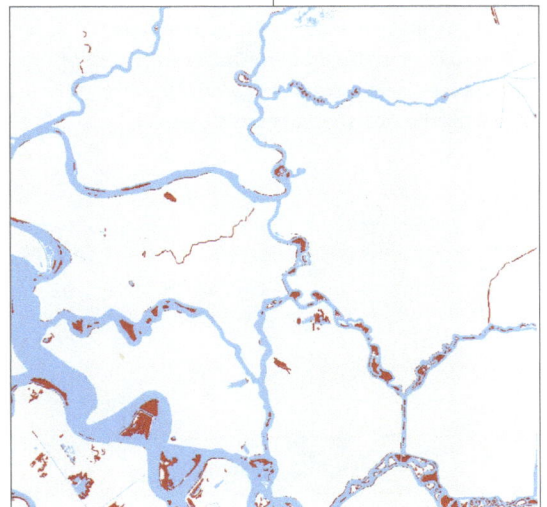

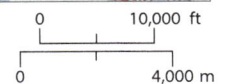

Liberty Island

Sherman Island

N

| 0 | 5 | 10 | miles |
| 0 | 10 | 20 km | |

1:575,000

| 0 | | 10,000 ft |
| 0 | | 4,000 m |

Existing marshes have little core habitat

This configuration leaves marsh wildlife vulnerable to edge effects

Large areas of core habitat are necessary for sustaining resilient marsh wildlife populations. While edges can be beneficial for wildlife populations—providing transition zones and a diversity of habitats—an increased edge-to-area ratio can increase predation and limit the value of core marsh areas. Historically, the marsh-channel edge was an important zone of exchange for both marsh and aquatic wildlife. Today, the small, isolated marsh patches with vastly altered hydrology have much less influence on the surrounding aquatic habitat. Modern marsh patches have little core area due to their small size and high edge-to-area ratios. Historically, 93% of the marsh was core habitat (>50 m from the marsh edge), while only 19% of marsh is core habitat today. Core areas experience different abiotic conditions, are less accessible to many predators, and are buffered from human disturbance in the modern landscape.[19] Fragmentation and development have increased the relative amount of edge habitat, although the absolute amount of marsh edge habitat in the Delta has been reduced. The character of the marsh edge has also been dramatically altered at both the upland and aquatic interfaces.

Historical (right) and modern (far right) extents of core marsh area. Historically, marsh edges transitioned to a tidal channel, riparian forest, seasonal wetland, or upland patch. In the modern Delta, edges are often steep levees, and account for proportionally more area than they did in historical marsh patches.

Methods: Marsh Core vs. Edge Area
Core area was defined as at least 50 m from the outside edge of the marsh.

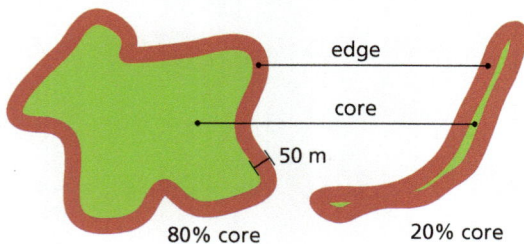

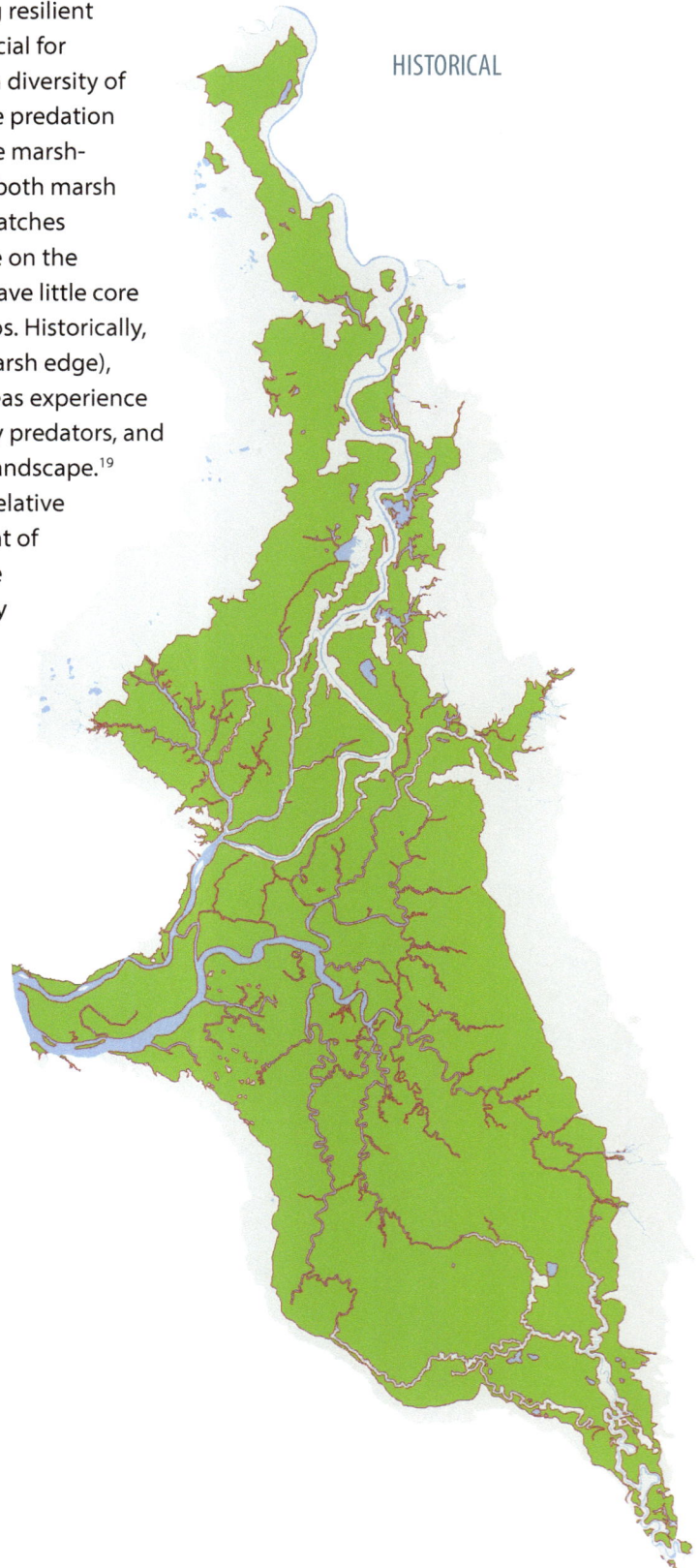

edge

core

50 m

80% core

20% core

HISTORICAL

MODERN

		Total area (ha)	
		Historical	*Modern*
■	Core	179,504	815
■	Edge	13,720	3,522
	TOTAL	**193,224**	**4,338**

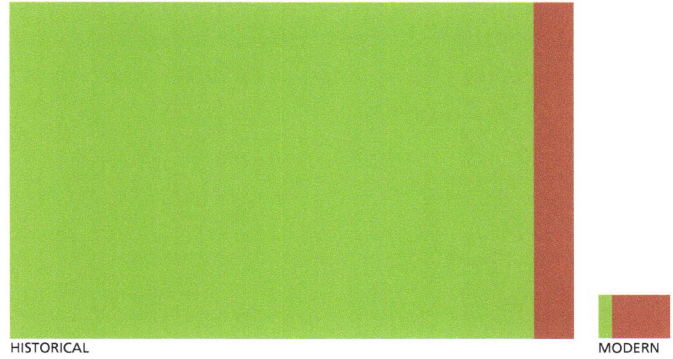

HISTORICAL   MODERN

Today's marsh habitat is mostly edge (left and above). Over 80% of existing marsh is edge area. The two large marsh patches at Liberty Island are 38% and 56% core area. The large marsh patch at Sherman Island is 34% core area. In the modern Delta, edges are often hard structures such as levees, roads, or agricultural land uses. There has been a 99.5% decline in core marsh area.

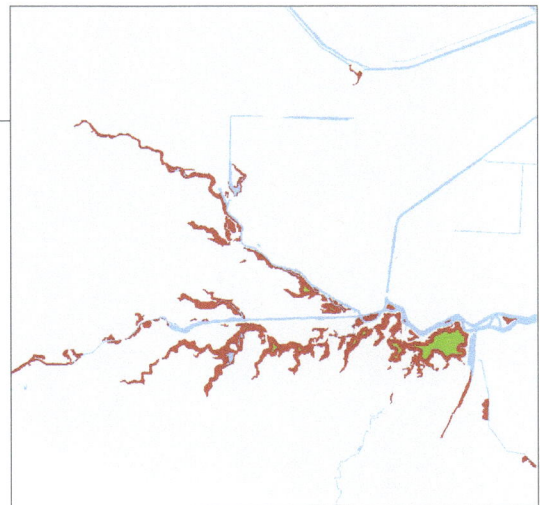

Liberty Island

Lindsey Slough

Sherman Island

0   3,000 ft

0   1,000 m

N

0   5   10 miles

0   10   20 km

1:575,000

Lindsey Slough (above) is one of the largest remaining patches in the Delta (second only to Liberty and Sherman islands), but its long linear shape (14% core area) makes this marsh vulnerable to edge effects.

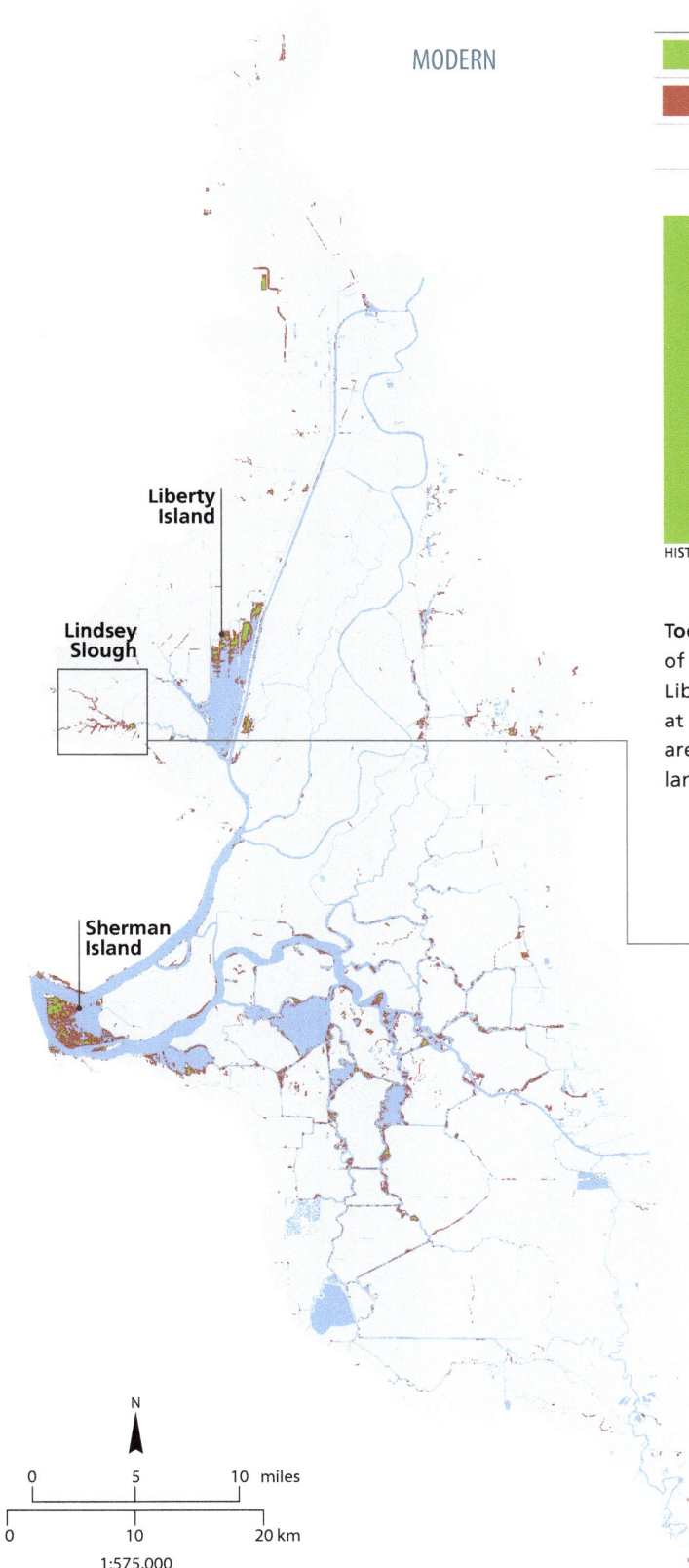

There are no modern analogues to the historical large, complex marshes

Even the highest quality remaining marsh patches are highly modified

Fragmentation has decreased the value of marsh to wildlife by reducing the size of marsh patches available, reducing the connectedness between marsh patches, and increasing edge effects. Areas of highest value to marsh wildlife are areas of core habitat that are either within large marsh patches (>100 ha) or are within small patches less than 1 km from a large marsh patch. Nearly all of historical marsh in the Delta met this criteria historically (179,495 ha or 93%), while only 491 ha (0.25% of the historical marsh area and 11% of the modern marsh area) meets this criteria today—a **99.7%** reduction in the extent of high quality habitat.

Methods: High value marshes

Combining the previous metrics, we can define areas of highest value to marsh wildlife as areas of core habitat that are either within large marsh patches (>100 ha) or are within smaller patches that are near (<1 km) large marsh patches.

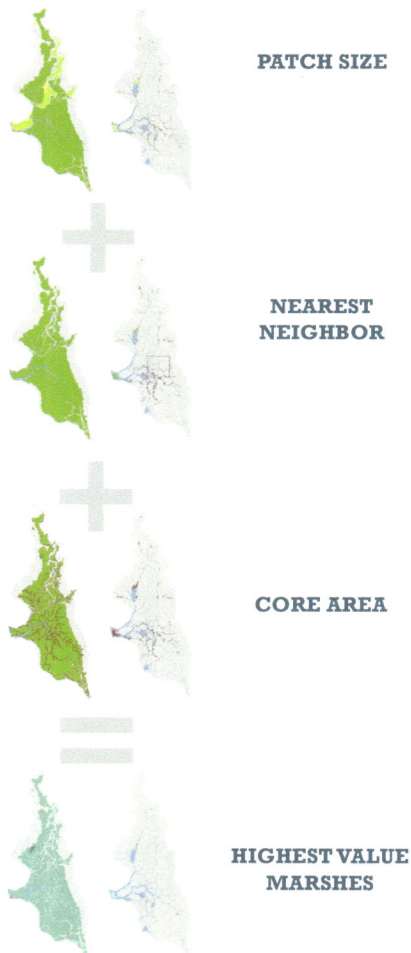

PATCH SIZE

NEAREST NEIGHBOR

CORE AREA

HIGHEST VALUE MARSHES

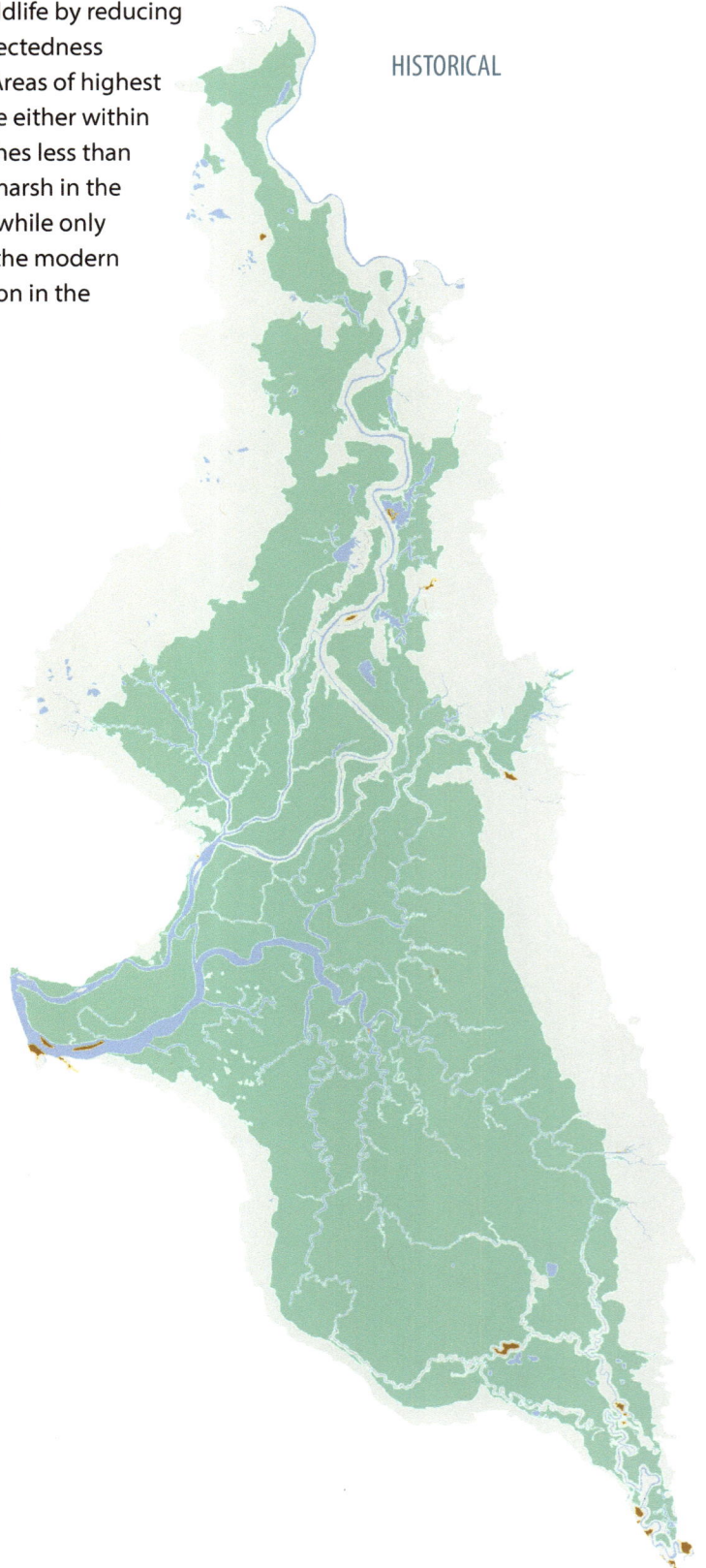

HISTORICAL

The largest current marsh patches are at Sherman and Liberty islands **(below)**. Marshes in both areas were recently formed as a result of levee failures and dredge disposal. The plant communities and channel networks at these sites differ from historical conditions as a result of the site histories.

MODERN

	CORE HABITAT	EDGE HABITAT	**Fragmentation index for historical and modern maps**
			Large (>100 ha) marsh patches (more likely to sustain persistent populations)
			Marsh near (<1 km) to large marsh patches (areas more likely to be recolonized and maintain gene flow)

Liberty Island marsh (right) formed after an accidental levee failure in 1998. This marsh, which is adjacent to the Yolo bypass, is bisected by levees and artificial channels, making it very different from the tule-dominated marsh native to this area that had very few channels. However, Liberty Island displays important features of the historical Delta, including high residence time, high suspended sediment concentration, and large marsh area.

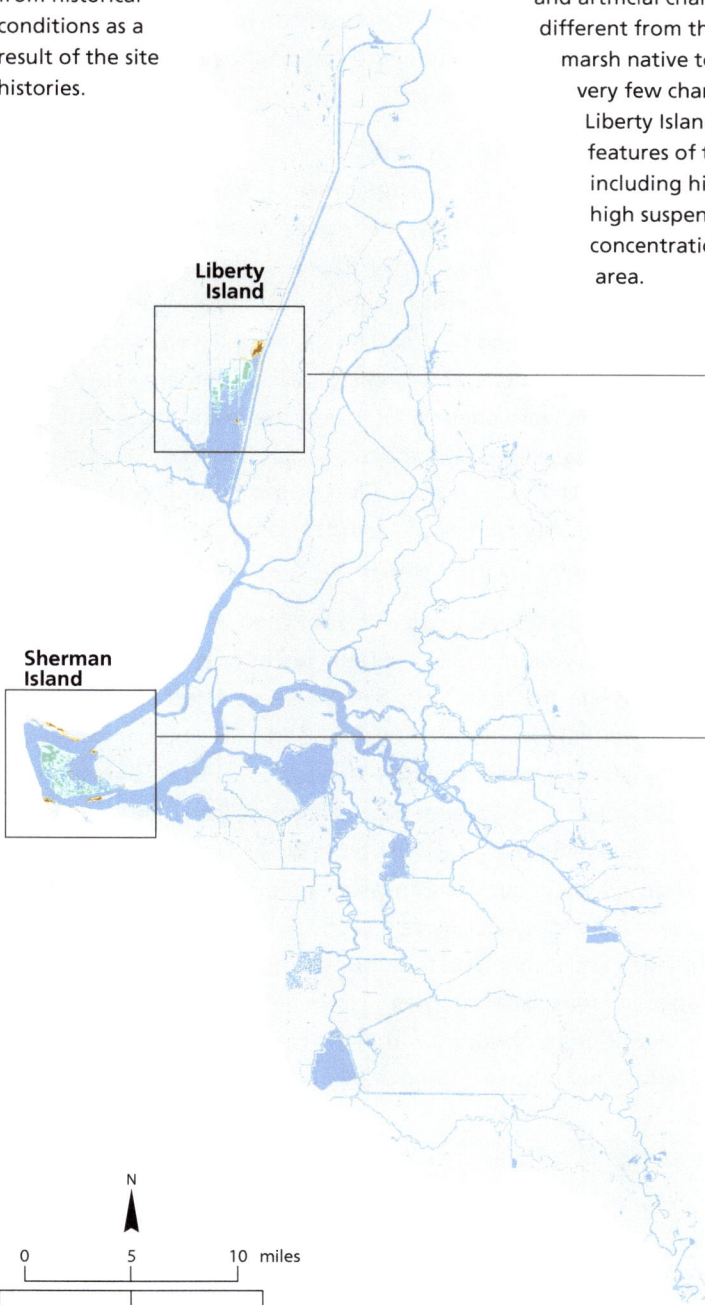

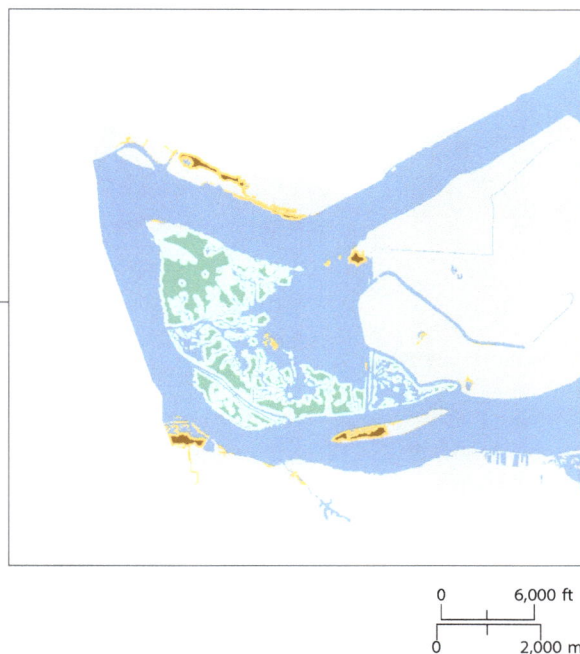

Liberty Island

Sherman Island

0   6,000 ft
0   2,000 m

0   6,000 ft
0   2,000 m

N

0   5   10 miles

0   10   20 km

1:575,000

Sherman Island (above) was flooded when its levees failed in the early 1900s. This marsh sits at the brackish (saltier) extreme of the salinity gradient within the Delta. The site was used for the deposition of dredge spoils until the 1960s.[20] Although it is large compared to other patches, it has lots of aquatic edge surrounded by invasive aquatic plants, has few distinct marsh channels, and sits at a low elevation in the tidal frame.

6. Life-History Support for Waterbirds

The historical Delta was important to many species of waterbirds, and supported both wintering and breeding birds. Waterbirds in the Delta included ducks, geese, swans, shorebirds, grebes, cormorants, bitterns, egrets, herons, ibises, rails, and terns.[1] The wetlands of the Central Valley, including the Delta, were associated with extraordinarily high concentrations of wintering waterfowl (ducks, geese, and swans). These wetlands also supported smaller but significant populations of breeding waterfowl, particularly dabbling ducks. The Delta provided year-round support to herons, egrets, and cormorants that nested and roosted in riparian trees and foraged in the extensive adjacent marshes. Coots, moorhens, and grebes likely inhabited the marshes and open waters of the Delta year-round, and Forster's Terns and Black Terns likely nested within the marsh. Waterbird species, such as cranes and shorebirds, which now rely on managed wetlands and flooded agricultural fields, likely took advantage of suitable habitats in the historical Delta, although their exact historical habitat associations are unclear.[2]

Large numbers of waterfowl—an estimated 35-50 million birds—overwintered in the Central Valley historically.[3] This area was a key stopover along the Pacific Flyway, a north-south migration route of global importance for waterfowl and other birds. While the relative value of the Delta among these Central Valley wetlands is unclear, reports from early explorers attest to the abundance of waterfowl within the Delta.[4] Migratory waterbirds adapt to changes in the landscape at a large scale, so the relative importance of Suisun, the Delta, and the Central Valley may have varied over time in response to changes in weather, water conditions, and food availability.[5] Modern waterfowl management focuses on the importance of seasonal wetlands because of the relative abundance of moist-soil seeds in these habitats compared to permanently flooded and tidal wetlands.[6] However, historically the low seed density in tidal wetlands may have been offset by the extensive acreage, leading to high total seed abundance.[7] Other food resources, including rhizomes, may also have been more important to wintering waterfowl using the historical Delta.

Different species of wintering waterfowl likely keyed in on different food resources and habitats within the Delta. Wintering waterfowl common in the Delta historically included Tundra Swans, Snow Geese, Ross' Geese, Greater White-fronted Geese, Canada Geese, Northern Pintails, Mallards, American Wigeons, Green-winged Teals, Northern Shovelers, Gadwalls, and Canvasbacks.[8] Emergent aquatic plants, submerged aquatic vegetation, moist-soil seeds, and invertebrates were all important food sources to these waterfowl.[9] Water depth in channels, lakes, and ponds determined which species could forage most efficiently, with dabbling ducks such as Northern Pintail preferring shallower water and diving ducks such as Canvasback preferring deeper channels and ponds. Swans foraged primarily on submerged aquatic vegetation, while geese grazed in seasonal wetlands and adjacent uplands and also fed on tuberous plants in wetter areas.[10] Waterfowl were unlikely to have foraged in areas of dense tules.[11] However seasonal and perennial lakes within the Delta, along with smaller ponds embedded within the marsh, were known to have supported high densities of waterfowl historically.[12] Regular disturbance via flooding, wildlife

Waterbirds (below).
Left to right:
Northern Pintail,
Sandhill Cranes at
Merced NWR, geese
migration, Mallards.

Photo Credits: Unknown,
USFWS; Lee Eastman,
USFWS; Unknown, USFWS;
Scott Flaherty, USFWS

wallows, and burning helped maintain these open water habitats. Geese themselves helped to maintain these ponds by clearing large areas of aquatic vegetation.[13]

Migrating shorebirds using the Pacific Flyway also undoubtedly took advantage of wetland habitats within the Delta, although records of particular species are lacking. Some shorebirds may also have bred in the Delta. Habitats frequently used by shorebirds in the modern Delta (e.g., wastewater treatment facilities, agricultural fields) are without historical equivalents. The sparse extent of mudflats in the central Delta historically, in contrast to the neighboring San Francisco Bay, would have limited shorebird use there. Shorebirds likely took advantage of what mudflat was available in the Cache Slough area and the short-statured vegetation in wet meadows, seasonal wetlands, and grasslands along the periphery of the Delta.[14] Available shorebird habitat historically would have shifted in time and space as water levels changed. Curlews and ibises likely foraged in grasslands and vernal pools.[15] Sandhill Cranes typically forage in low vegetation lacking shrubs and trees that might block their view of predators, and may have also used these habitats.[16] Avocets and stilts may have nested in the Delta, particularly in areas of marsh dominated by low rushes and grasses in the south Delta.[17]

The Central Valley was an important area for breeding waterbirds historically. Duck species that bred in the Delta included Mallards, Gadwalls, Cinnamon Teals, Northern Pintails, and possibly Redheads and Canvasback.[18] Upland areas adjacent to the marsh, or higher areas (like beaver and sand mounds within the marsh), offered nesting opportunities above flood waters.[19] Areas of open water within or adjacent to freshwater emergent marsh were used as brooding habitat for young birds. Waterfowl may have moved a considerable distance between nesting and brooding sites, particularly when nesting occurred in brackish areas, such as Suisun Marsh.[20] Freshwater marshes were also important post-breeding sites for molting birds. Many species of waterfowl molt all their primary flight feathers simultaneously, rendering them temporarily flightless, and the tall, dense vegetation in these wetlands provided critical cover to these vulnerable birds.[21] Riparian forests provided nesting opportunities for cavity-nesting Wood Ducks and supported large rookeries of herons, egrets, and cormorants.[22]

The modern Delta provides less support for waterbirds due to the extensive loss of wetland habitat. Managed wetlands and agricultural fields are key components of the modern landscape for both wintering and breeding waterbirds, as natural wetlands no longer provide adequate space or food resources for wintering waterfowl.[23] Although these managed habitats are now crucial for waterbirds they differ from historical wetland habitats in several important ways. Grain crops provide food resources that are carbohydrate-rich but sometimes nutrient-poor, and these areas lack the invertebrate communities important for particular species at certain times of year.[24] Management is often focused on supporting particular threatened and endangered species, such as Sandhill Cranes, or supporting economic interests, such as duck clubs, and may provide less support for non-target species. Water quality within flooded agricultural fields can be affected by fertilizers and pesticides. In addition, some modern waterfowl habitat may be increasingly threatened by levee failure or water shortages.[25] Restoring wetlands has the potential to shift waterbird support back to natural areas. This support is likely dependent on the size of restored areas.

Delta wetlands supported large numbers of waterbirds historically

Variable conditions in the Delta provided support for different waterbirds in different seasons

Different waterbirds species were able to take advantage of different parts of the Delta throughout the year. Waterfowl wintered in the Delta in particularly high numbers.[26] The combination of vegetative structure and flooding patterns determined habitat suitability for different waterbird species. There was a high degree of variability in these habitats, with the amount and location of suitable habitat changing significantly from one year to the next. Waterbirds responded to landscape heterogeneity on a finer scale than can be mapped here. For example, wintering waterfowl were found in high densities in small ponded areas within freshwater marshes. The abundant food resources and protection from predators were due in part to the large size of the Delta wetlands. Although waterfowl food density in the Delta wetlands may have been lower than in intensively managed freshwater marshes today, the large historical extent of marshland likely resulted in overall food supplies that rivaled or exceeded the modern-day support.[27] The remaining natural marshes and seasonal wetlands are too small to support abundant waterbirds, and thus in the modern Delta managed wetlands and agricultural fields provide most of the habitat value for waterbirds.

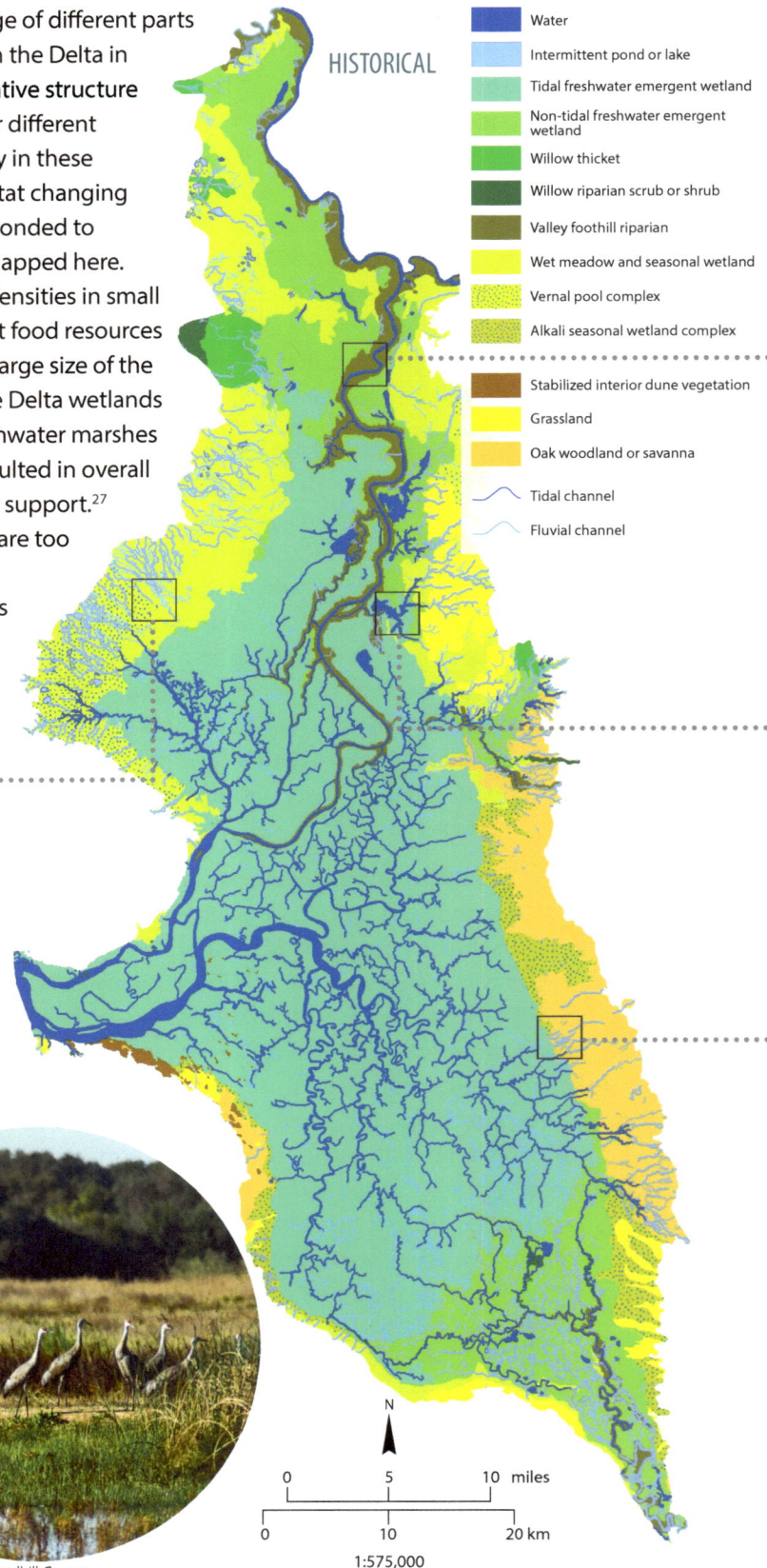

HISTORICAL

	Water
	Intermittent pond or lake
	Tidal freshwater emergent wetland
	Non-tidal freshwater emergent wetland
	Willow thicket
	Willow riparian scrub or shrub
	Valley foothill riparian
	Wet meadow and seasonal wetland
	Vernal pool complex
	Alkali seasonal wetland complex
	Stabilized interior dune vegetation
	Grassland
	Oak woodland or savanna
	Tidal channel
	Fluvial channel

SANDHILL CRANES

The Delta was an important wintering location for Sandhill Cranes. Areas of shallow water and short vegetation likely provided good roosting areas with easy detection of predators. Seasonal wetlands around the edge of the Delta offered this type of habitat historically.[28]

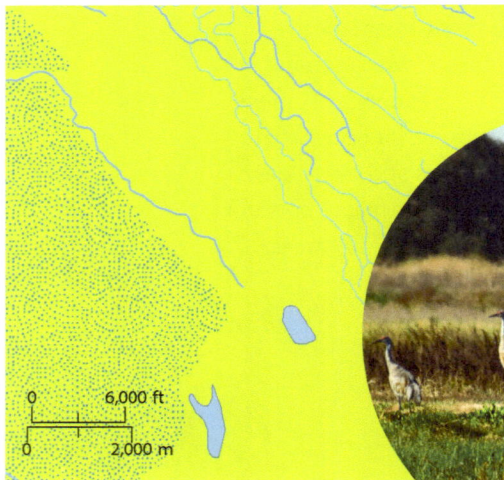

0   6,000 ft
0   2,000 m

N

0   5   10   miles

0   10   20 km

1:575,000

Sandhill Cranes

Photo Credits (left to right): Justine Belson, USFWS; Linda Wanczyk, SFEI; Charles Sharp, Wikipedia Commons; Dan Cox, USFWS

COLONIAL ROOSTING BIRDS

Colonies of herons, egrets, and cormorants used large trees in the riparian forests of the north Delta for roosting and nesting.[29]

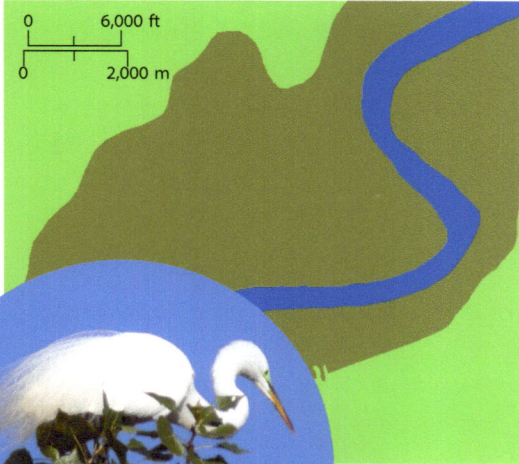

Roosting heron

WINTERING WATERFOWL

Wintering waterfowl congregated in large numbers in areas of open water within freshwater marsh. Common species included Northern Pintails, Snow Geese, Ross' Geese, and Tundra Swans. Seeds and tubers of marsh plants were particularly important food resources for these species.[30]

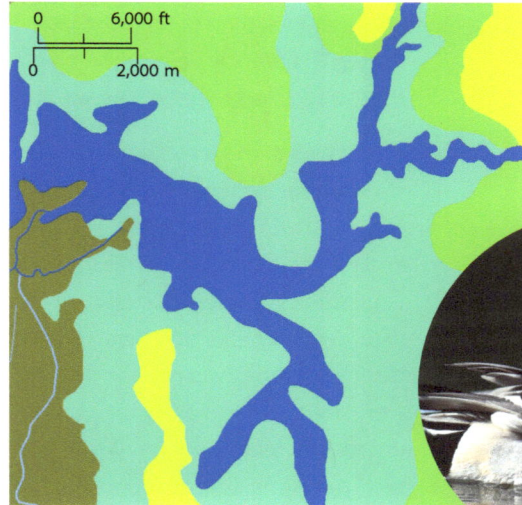

Northern Pintail

BREEDING DUCKS

Several species of dabbling ducks, including Mallards and Gadwalls, bred in the Delta in significant numbers. Areas of higher elevation, above flood waters, were critical for nesting. Areas of open water with nearby vegetative cover were needed for brooding birds.[31]

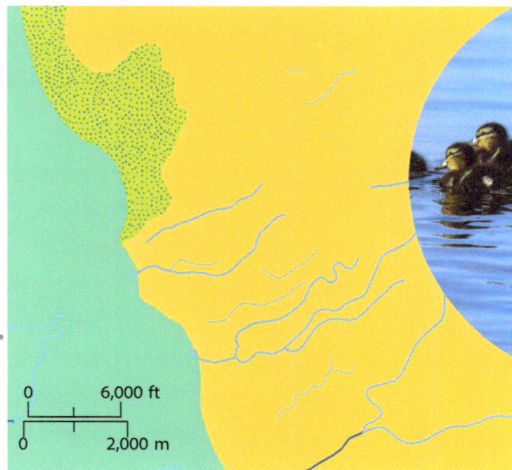

Mallard ducks

Flooding in the historical Delta (below). The diagram below relates flooding in the historical Delta (see **details and legend on page 40**) with patterns of waterfowl and shorebird use.[32]

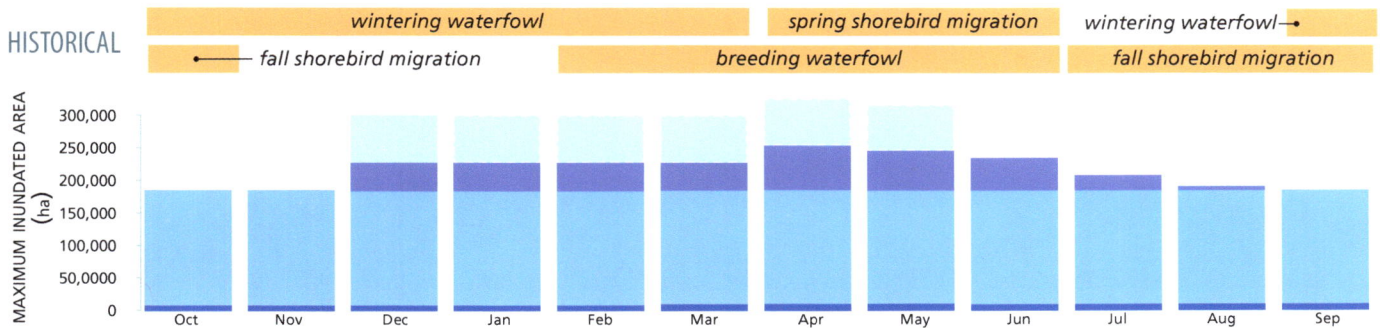

HISTORICAL

wintering waterfowl | spring shorebird migration | wintering waterfowl→
fall shorebird migration | breeding waterfowl | fall shorebird migration

MAXIMUM INUNDATED AREA (ha)

Oct | Nov | Dec | Jan | Feb | Mar | Apr | May | Jun | Jul | Aug | Sep

7. Life-History Support for Riparian Wildlife

Woody riparian habitats form the interface between aquatic environments and adjacent areas, providing structurally complex environments that support diverse species. Historically, broad riparian forests and willow shrubs, elevated on natural levees, lined the Sacramento and San Joaquin rivers and their major tributaries. These habitat types were shaped by hydrologic and geomorphic disturbance: floods built up natural levees and stimulated successional processes of riparian forests. These natural levees extended far into the marsh, providing dryland access deep into the Delta's marshes for terrestrial species.[1] The vertical structure and plant diversity of riparian forests provided abundant food resources and sites for numerous resident and migratory birds to forage, nest, and roost.[2] The woody vegetation also provided shade and contributed allochthonous inputs to the river that supported aquatic species, including anadromous fish.[3]

There was considerable heterogeneity within woody riparian habitats, particularly between riparian forests in the north and south Delta. Riparian forests historically were largely confined to the north and south Delta because of the Sacramento and San Joaquin rivers' loss of stream power and ability to build large natural levees as they entered the central Delta. In the north Delta, riparian vegetation consisted of broad riparian forests dominated by oaks and sycamores, often a half mile wide, with a multilayered and diverse understory composed of willow, alder, buttonbush, dogwood, box elder, buckeye, grape, wild rose, and numerous herbaceous species. Riparian areas along the San Joaquin River were narrower and dominated by willows and other shrubs. There was considerable lateral and upstream/downstream heterogeneity within these habitats. Vegetation varied with the elevation of natural levees, with the highest areas supporting large trees, while the wetland and channel edges supported willows and grasses. Compared with areas farther upstream, the downstream reaches of woody riparian habitats were narrower and increasingly dominated by willows and marsh vegetation. Vegetative structure was influenced by channel size, with larger channels often supporting more extensive woody riparian habitat, due to the larger size of their natural levees. Willow-fern complexes in the central Delta may have also provided some support to riparian species, though they differed in habitat structure and continuity from other riparian habitat types.[4]

Despite comprising only a small proportion of the total area of the historical Delta (7%), riparian forests provided important habitat for a diverse suite of species. Woody riparian habitats likely served as movement corridors for far-ranging terrestrial mammals such as coyotes and mule deer as well as smaller mammals including gray fox, long-tailed weasels, and ringtails.[5] The south Delta forests provided important habitat for several endemic species, including the riparian brush rabbit, riparian woodrat, and valley elderberry longhorn beetle.[6] Riparian forests in the Central Valley were particularly important to both resident and migratory birds, supporting a diverse and abundant assemblage of species.[7] These forests contained high densities of breeding birds compared to other habitats, and provided nesting habitat for Red-shouldered Hawks, Swainson's Hawk, Western Yellow-billed Cuckoos, Willow Flycatchers, Least Bell's Vireos, Yellow Warblers, Yellow-breasted Chats, and Blue Grosbeaks.[8]

Riparian forests offered many nesting niches—on the ground, in shrubs and trees, on branches, and in tree cavities. Forests dominated by large oaks and sycamores were particularly important to cavity nesters, including Wood Ducks, Downy Woodpeckers, Oak Titmouse, and Ash-throated Flycatchers. Large riparian trees supported breeding and roosting colonies of herons, egrets, and cormorants. Oak-dominated riparian habitat supported high densities of wintering birds, especially Sharp-shinned Hawks, Hermit Thrushes, Yellow-rumped Warblers, and Golden-crowned Sparrows.[9] These habitats were also used by passing migrants, and may have been especially important to fall migrants that glean insects (e.g., Wilson's Warblers, Western Tanagers) because other green, insect-rich vegetation was sparse at that time of year.[10]

Existing woody riparian habitat occupies 40% of its historical extent, but these areas are now severely fragmented, with virtually no wide corridors of riparian forest remaining. Today's narrow patches are structurally simpler and more homogeneous than historical woody riparian habitats, often lacking the complex microtopography, moisture gradients, vegetative structure, and diversity which provided essential ecosystem services, such as erosion control and riparian forest regeneration.[11] As mapped, 90% of historical Delta woody riparian habitat was riparian forest; today only 58% is forest, and the rest is willow shrub habitat.

Riparian species once common in the Delta are in decline. The endangered Western Yellow-billed Cuckoo, Least Bell's Vireo, and other species no longer breed in the Delta.[12] The decline in nesting Cooper's Hawks and Western Yellow-billed Cuckoos is thought to be a direct result of the loss and fragmentation of available habitat, as both species require large territories to breed.[13] Riparian species have been impacted by degraded habitat quality, that is often hydrologically disconnected from adjoining rivers. Agricultural development adjacent to woody riparian habitats has facilitated movement of non-native Brown-headed Cowbirds and European Starlings into these habitats, negatively impacting native birds through nesting cavity competition and reduced nest success.[14] Levees (with hardened edges and lack of regeneration from flooding) adjacent to woody riparian habitats have allowed non-native predators (**feral dogs, cats, and rats**) increased access to these habitats, to the detriment of riparian brush rabbits, riparian woodrats, and other species.[15] Riparian brush rabbits have also been impacted by the lack of suitable habitat above regular flood levels that previously provided protection from weather and predators.[16]

The position of woody riparian habitats within the modern Delta landscape has become less coherent. Whereas woody riparian habitat historically lined large rivers and tributaries in continuous bands, today small disconnected riparian patches exist scattershot across the entire Delta, including the central Delta where these habitats were historically absent. In many instances "riparian habitats" are separated from the rivers that created them by artificial levees and upland areas, and are thus disconnected from the physical processes sustain them. Restoration of continuous, self-sustaining woody riparian habitats in the Delta may be particularly important in the face of climate change, because these habitats provide linear habitat connectivity, link aquatic and terrestrial ecosystems, and create thermal refugia for wildlife.[17]

Riparian wildlife (below). Left to right: riparian brush rabbit, riparian vegetation, long-tailed weasel, male valley elderberry longhorn beetle, coyote.

Photo Credits: Brian Hansen, USFWS; William Miller; Rick Kimble, USFWS; Jon Katz and Joe Silveira, USFWS; Steve Thompson, USFWS

Modern woody riparian habitat is highly fragmented

Large, continuous riparian forest is gone, except along the Cosumnes River

The woody riparian habitats in the Delta today are severely reduced, fragmented, and degraded. Historically, woody riparian habitat existed as large continuous corridors along the major Delta rivers and tributaries in the north and south Delta. Modern woody riparian habitat is a scattering of small discontinuous patches throughout the Delta that no longer support resident and migratory species to the same degree, due to differences in habitat quantity, quality, and landscape configuration. Historical gallery riparian forests in the north Delta had canopies of oak and sycamore with a complex understory of alder, willow, blackberry, and many other species. Modern woody riparian habitats are smaller, simpler systems, largely dominated by willow and invasive understory plants associated with narrow levees, and are not as exposed to regenerative disturbance regimes. Small habitat fragments support fewer species and smaller populations with more edge effects. Researchers found, for example, that Western Yellow-billed Cuckoos in northern and southern California are six times more likely to be found in habitat patches 40-80 ha than patches 20-40 ha (they were detected in all habitat patches larger than 80 ha).[18]

Western Yellow-billed Cuckoo

Historical riparian habitat was predominately continuous forest (right), while today woody riparian habitat is scattered throughout the Delta in small isolated patches (far right). The longest stretch of contiguous riparian forest[19] historically spanned more than 55 km (from the Feather River confluence to Miner Slough), providing a migration corridor across much of the Delta. The longest current stretch of woody riparian habitat extends 16 km (along Elk/Sutter Slough).

Methods: Riparian patch size

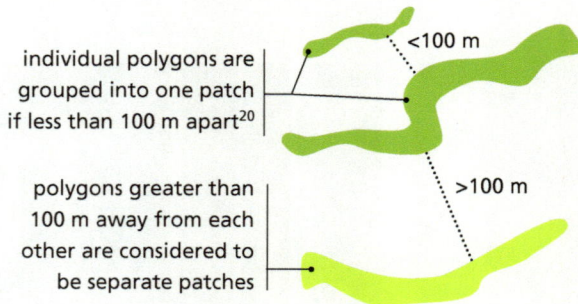

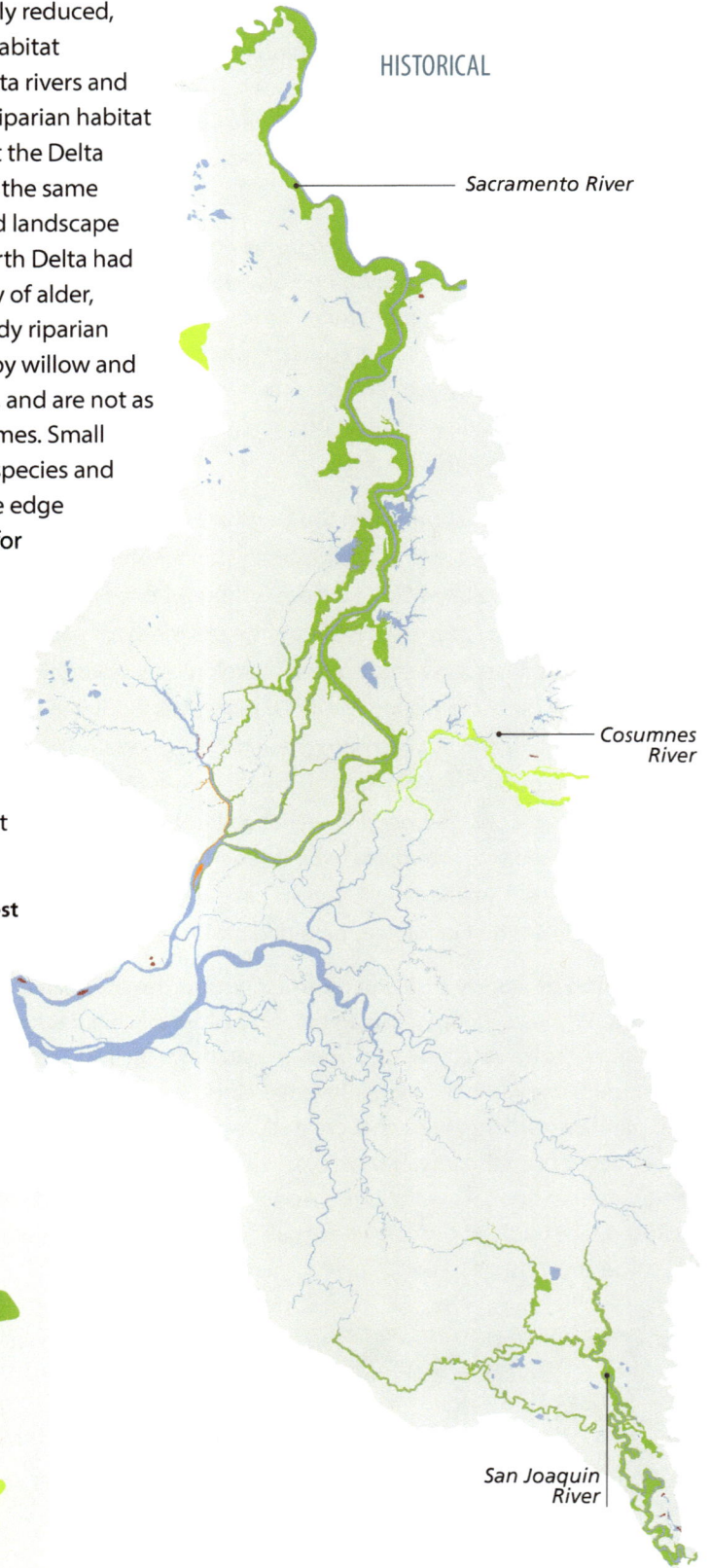

individual polygons are grouped into one patch if less than 100 m apart[20]

<100 m

polygons greater than 100 m away from each other are considered to be separate patches

>100 m

HISTORICAL

Sacramento River

Cosumnes River

San Joaquin River

Photo Credit: Factumquintus, Wikipedia Commons

MODERN

Sacramento River

Cosumnes River

San Joaquin River

Woody riparian patch size class (ha)	Total area (ha)	
	Historical	*Modern*
<= 20 ha	88	1,991
20 - 80	113	1,364
80 - 320	0	1,470
320 - 1,280	1,594	2,066
>1,280	15,449	0
TOTAL	**17,244**	**6,890**

HISTORICAL   MODERN

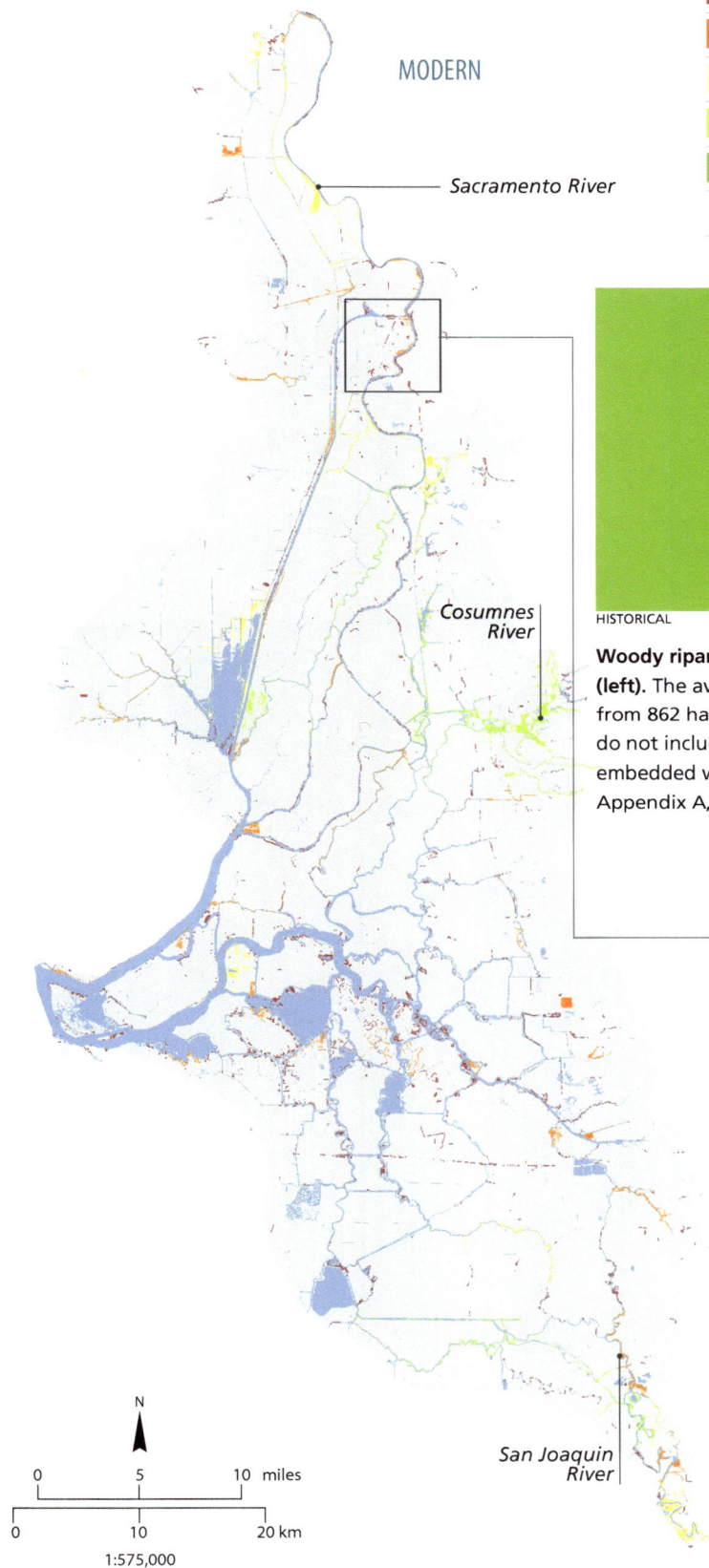

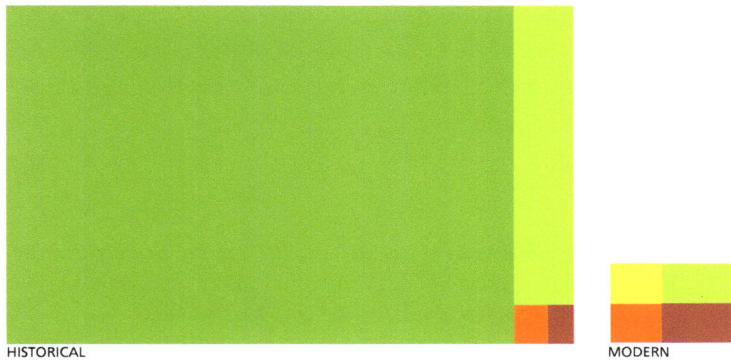

Woody riparian habitat extent in the Delta has been reduced by 60% (left). The average woody riparian patch size in the Delta has decreased from 862 ha (SD = 2,785 ha) to 6 ha (SD = 45 ha). These maps and figures do not include an estimated 3,500 ha of unmapped willow patches embedded within the tule marshes of the historical central Delta (see Appendix A, page 91).

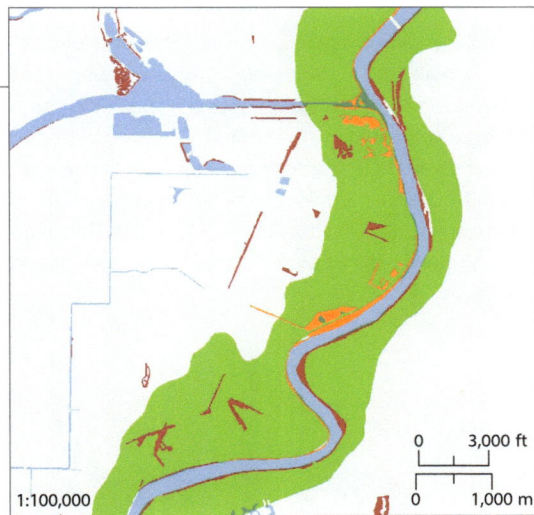

1:100,000   0   3,000 ft   0   1,000 m

Historically, woody riparian habitat along the Sacramento River formed a wide continuous band of forest—the woody riparian habitat that exists today is made up of many small habitat fragments (above). Historical woody riparian habitat is shown in green with modern woody riparian habitat overlaid in red and orange (depending on their patch size).

N

0   5   10 miles

0   10   20 km

1:575,000

Wide woody riparian habitat has declined by 72%

Wide riparian corridors provided habitat complexity and supported species with large home ranges

Historically, the Delta contained wide riparian corridors, particularly in the north Delta where the riparian forest could exceed a mile in width. These wide riparian corridors supported complex habitats, with many vegetative zones influenced by elevation, moisture gradients, and disturbance patterns. Interior woody riparian habitats were buffered from edge effects and supported species that need large riparian areas, particularly nesting Western Yellow-billed Cuckoos and Cooper's Hawks, as well as far-ranging mammals, such as coyotes. A review of the literature on the effect of riparian width on birds found that while a riparian width of 100 m was sufficient for many species, a width of 500 m was necessary to support the complete avian community.[21]

Wide woody riparian habitat in the historical (right) and modern (far right) Delta. Woody riparian habitat wider than 500 m is shown in dark green, and habitat wider than 100 m is shown in light green. The width of the riparian habitat was determined by the river's ability to build natural levees above the marsh of the interior Delta, creating well-drained soils that supported trees. In general, the width and height of riparian habitat declined as the large river systems spread into the central Delta.

Methods: Calculating riparian widths

Transects were cast perpendicular to channels at 100 m intervals. The width of each transect was summed (excluding channels) where it overlapped riparian habitat (shown in yellow below). On the map and in the diagram below, transects wider than 100 m are in light green, and transects wider than 500 m are in dark green. Transects less than 100 m wide (dotted lines below) are not shown on the map. See Appendix A for details.

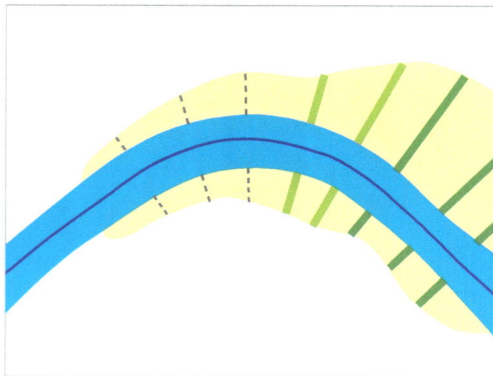

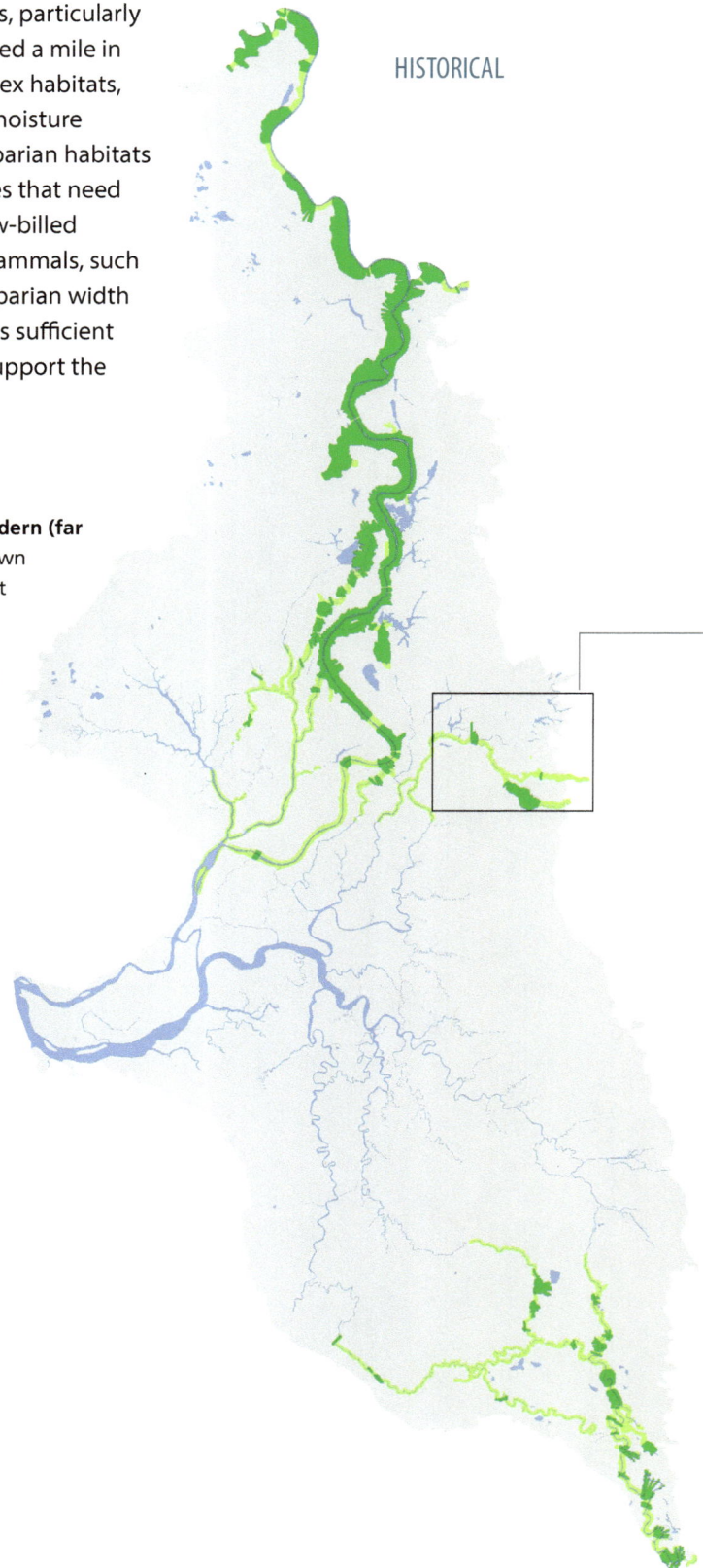

HISTORICAL

MODERN

	Woody riparian habitat width (m)	Woody riparian habitat length (km)	
		Historical	*Modern*
NOT SHOWN ON MAP	0 - 100 m	37	626
	100 - 500	239	87
	> 500	116	11
	TOTAL	**393**	**723**

The length of wide riparian habitat has declined. Nearly all (87%) of the woody riparian habitat in the modern Delta is less than 100 m wide, and less than 2% of the habitat is greater than 500 m wide.

The largest cluster of woody riparian habitat >500 m wide is the restored habitat along the Cosumnes River (below). Historically, the Mokelumne River supported wide riparian forest on natural levees. Due to changes in groundwater levels, land use, and channel incision, the majority of woody riparian habitat is now located along the Cosumnes River in an area that historically supported freshwater emergent marsh and wet meadow.[22] However, this wide and complex riparian habitat provides many of the ecological functions that riparian habitat provided historically in other places in the Delta.

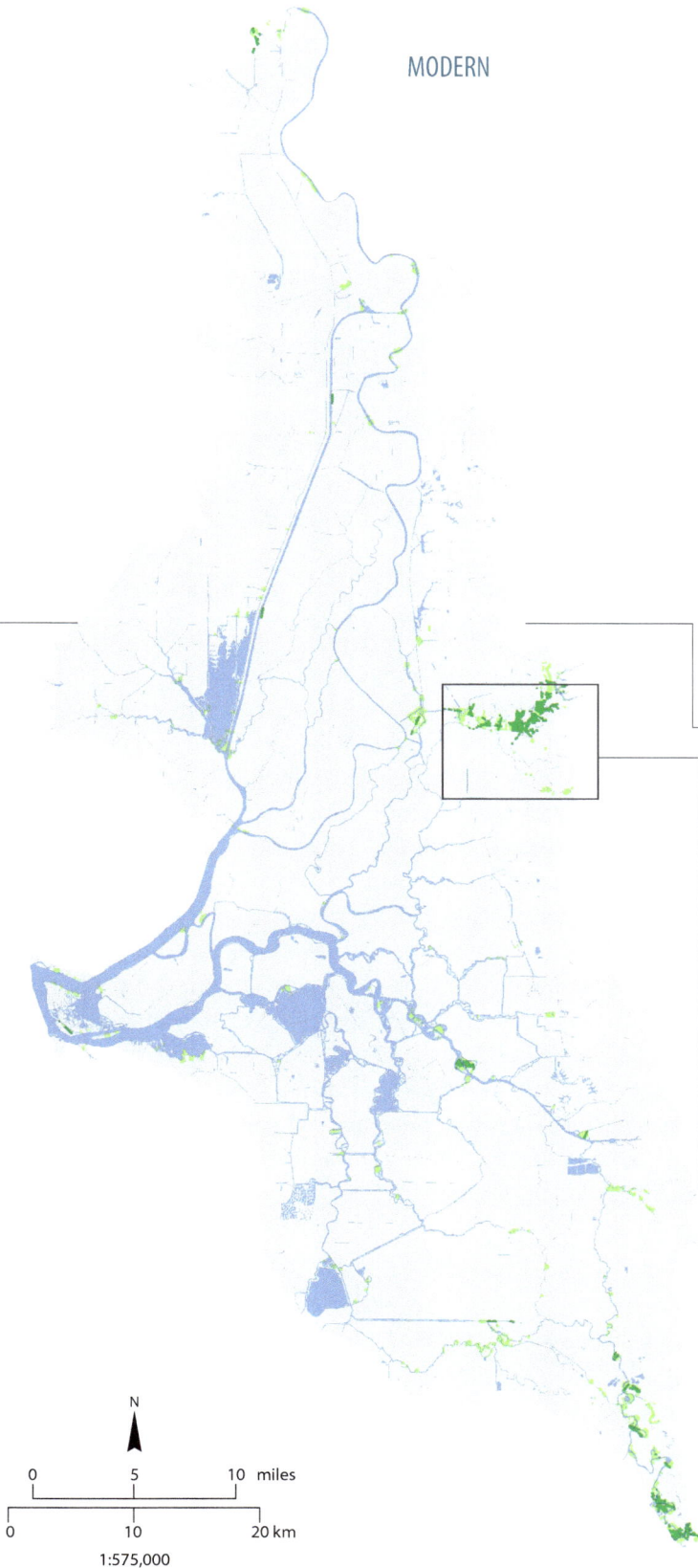

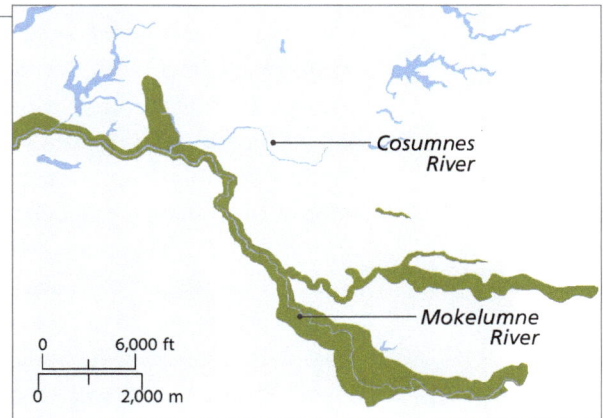

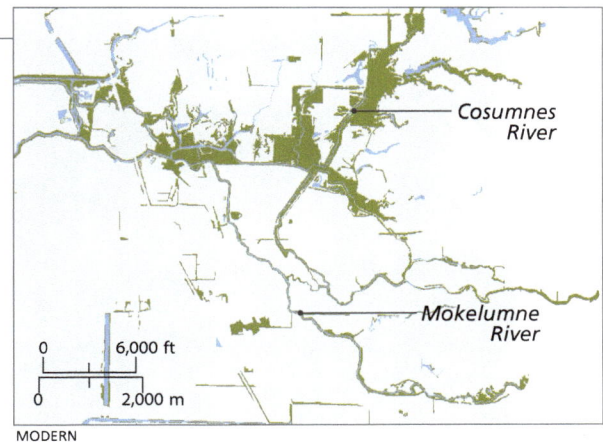

Cosumnes River

Mokelumne River

0   6,000 ft

0   2,000 m

HISTORICAL

Cosumnes River

Mokelumne River

0   6,000 ft

0   2,000 m

MODERN

N

0   5   10   miles

0   10   20 km

1:575,000

8. Life-History Support for Marsh-Terrestrial Transition Zone Wildlife

The edge (or transition zone) of the Delta marsh provided ecological functions critical for many wildlife groups. The ecological functions of this transition zone varied depending on its position within the Delta. The extensive freshwater emergent marsh was bounded by elevation, with the upslope side transitioning into terrestrial habitats across a broad zone. Seasonal wetlands, including alkali wetlands and vernal pools, were found along the gently sloping upland transition in the northwest and southwest Delta,[1] while grasslands, oak savannas, and woodlands were found along the steeper, well-drained alluvial fans bordering the Delta to the east. The transition zone occurred primarily along the periphery of the Delta, with the exceptions of long corridors of riparian forest extending into the marsh and scattered sand dunes that punctuated the marsh in the southwest Delta. The relatively continuous transition zone along the periphery of the Delta would have supported dispersal and other movement of amphibians and reptiles dependent on both wetland and upland habitats (e.g., giant garter snake, California red-legged frog, and Western pond turtle).[2] Riparian corridors provided predators like bats, weasels, and coyotes with access to abundant prey from the productive marsh.[3] Riparian habitat also provided North American river otters with denning sites near the marsh but above frequently flooded elevations. Sand dunes (isolated upland patches within the Delta) provided important flooding refuge and predator protection.[4] The central Delta consisted of tidal marsh channels that lacked the stream power to build large natural levees, leaving this part of the Delta farther from any terrestrial transition zone.

Habitats occurring next to the marsh varied across the Delta, based on gradients in hydrology, topography, and soil. Along the northwest Delta where slopes were gradual and characterized by heavy clay soils, the marsh transitioned to seasonal wetlands interspersed with vernal pools. These seasonal wetlands were variable and complex, with inundation and vegetation patterns sensitive to small-scale changes in hydrology and topography. Seasonal wetlands in the northwest Delta were inundated by intermittent streams that lost channel definition before reaching the marsh and sometimes by the large floods of the Sacramento River. Along the eastern edge of the Delta the marsh transitioned to alkali wetland and oak savanna. Alkali wetlands, characterized by evaporative salt residues, were found in areas inundated only by extreme flooding. The oak savanna occurred on the well-drained soils of the alluvial fans that bordered the eastern side of the Delta, built by the Calaveras and Mokelumne rivers. To the south where soils were shallower, alkali wetlands were interspersed with grassland habitats. The interior dune scrub found along the southwestern edge of the Delta was a relic of Pleistocene dunes. The width and complexity of the transition zone was greater in areas with more gradual slopes, particularly areas supporting seasonal wetland.[5] These gradual transitions allowed movement and adaptation for particular species along moisture and elevation gradients.

The habitats adjacent to the marsh were key for wildlife in their own right, in addition to the transition zone species they supported. While none of these habitat types were unique to the Delta periphery, their proximity to Delta wetlands benefited the species they supported (e.g., by providing access to freshwater in the summer). The number of different habitat types adjacent to Delta marshes augmented the overall biodiversity of the region. Many of the species once associated with habitats adjacent to marsh are species of concern or otherwise important to land managers within the Delta today. Riparian forests supported migratory songbirds and several protected species of small mammals (e.g., riparian woodrat, riparian brush rabbit).[6] Seasonal wetlands provided habitat for many species of migratory waterbirds and amphibians.[7] Alkali wetlands and vernal pools supported many endemic plants and invertebrates.[8] Grasslands were important to many species now extirpated or uncommon in the Delta, including large mammals, such as grizzly bears, pronghorn, and tule elk.[9] Vernal pools, alkali wetlands, grasslands, and sand dunes are discussed in more detail below because of the number of endemic species they supported and their importance to overall Delta biodiversity.

The terrestrial transition zone was comprised primarily of seasonal wetlands which expanded the availability of wetland and aquatic habitat at certain times of the year. The majority of seasonal wetlands were found bordering the north Delta and encompassed a diverse range of plant communities, perhaps owing to variable inundation frequencies, dry-season dessication, topographic complexity, soil types, and freshwater inputs or "sinks" from tributaries. Vernal pools and alkali complexes were often intergraded with the seasonal wetlands, particularly in the southern parts of the Delta margin where drier conditions promoted the accumulation of salts in soils. When flooded, seasonal wetlands provided connectivity for terrestrial species such the giant garter snake between the nutrient-rich Delta and the surrounding valley, as well as short-term foraging habitat for certain aquatic species.

Vernal pools and alkali seasonal wetlands in particular supported many unique species. Vernal pools tend to support endemic species uniquely adapted to their hydrology. These are ephemeral wetlands characterized by shallow depressions that are inundated for too long to support upland species, but not long enough to support aquatic species.[10] Many vernal pool plants are specially adapted annuals that grow quickly as the ponds dry. Several invertebrates and amphibians use these pools to breed, taking advantage of the lack of predatory fish. Special status species supported by vernal pools included California linderiella, conservancy fairy shrimp, longhorn fairy shrimp, midvalley fairy shrimp, vernal pool fairy shrimp, and vernal pool tadpole shrimp.[11] The alkali seasonal wetlands that characterized much of the periphery of the Delta were complex habitats made up of small brackish ponds, perennially wet alkali marsh, alkali sink scrub, and seasonally inundated alkali meadow.[12] These habitats supported many unique plant species adapted to alkaline conditions, including saltgrass, swamp grass, Delta button celery, popcorn flower, iodinebush, San Joaquin spearscale, and the now potentially extinct caper-fruited tropidocarpum.[13]

The Delta edge (below). Left to right: Suisun Marsh, San Joaquin kit fox, Guadalcanal Mitigation Site, tule elk.

Photo Credits: Daniel Burmester, CDFW; Carley Sweet, USFWS; Gena Lasko, CDFW; Steve Martarano, USFWS

Grasslands were important to a diverse suite of wildlife, many of which are now locally threatened or endangered. Prior to non-native annual grasses establishing dominance, these habitats were believed to be dominated by forbs, with some annual and perennial grasses intermixed.[14] Grassland and savanna habitats were important to far-ranging large mammals that occasionally ventured into the marsh, including grizzly bears, mule deer, and tule elk. These grasslands supported many species of burrowing animals, such as California ground squirrels, California voles, and San Joaquin kangaroo rats, which created topography and structure important to Western Burrowing Owls, giant garter snakes, spadefoot toads, and California tiger salamanders. Swainson's Hawks foraged in grasslands historically.[15] The Meadowlark, Short-eared Owl, Horned Lark, Savannah Sparrow, and San Joaquin kit fox were also associated with these grasslands.[16]

Scattered sand mounds—high points of glacial-age eolian dunes—rose above the marsh plain, adding supra-tidal topographic variation and habitat complexity to the flat terrain of the western Delta. The mounds supported numerous species of plants and animals that would have otherwise been unable to persist within the Delta's tidal environment, such as lupine, the special status Antioch Dunes evening primrose, the western wallflower (Contra Costa wallflower), the endangered Lange's metalmark butterfly, and even live oaks on certain dunes with a developed soil profile. Tule elk were observed to have used these sites as protected breeding and foraging habitat, since the mounds offered some protection from larger predators less likely to venture far into the marsh. These areas of high elevation were also used and sometimes augmented by the indigenous communities who lived in and around the Delta. Sand dunes, as well as large man-made mounds, or middens, were often occupied by village sites, as they were in close proximity to the rich abundance of food and resources provided by the Delta but were protected from daily tidal flooding.[17]

The marsh-terrestrial transition zone in the Delta has been dramatically reduced, fragmented, and degraded. This loss is largely due to the 97% reduction in marsh and the conversion of adjacent habitats to agriculture and development. Much of the remaining marsh occurs as islands in the central and west Delta, in places where the marsh-terrestrial transition zone was never present historically. The terrestrial boundary of modern marshes, where it does exist, is often characterized by an abrupt transition to upland or man-made structures, such as a steep, sparsely-vegetated rock levees and other inflexible edges that offer little in the way of cover, gradients, or habitat value. In addition, remaining marsh patches may no longer provide the same food subsidy to terrestrial species because of their greatly reduced size. The marsh-terrestrial transition zone once formed a complex but continuous band, predictable along hydrological and elevation gradients. That transition zone is now fragmented and disorganized, making it difficult for wildlife to anticipate resources available from the edge.

The terrestrial habitats that occur in the Delta today are largely disconnected from the marsh and from the processes that established and maintained these habitats historically. The dominant habitats in the modern Delta are grasslands and seasonal wetlands that occur in the center of the Delta as often as the periphery. The location of many of these habitats makes them particularly vulnerable to sea level rise. The hydrology of seasonal wetlands is heavily managed and disconnected from seasonal flooding patterns, and seasonal wetlands are now found where perennial wetlands once existed. Agricultural fields and ditches provide a limited portion of the natural functions provided by seasonal wetlands, do not support the same hydrologic regime, and experience stress from human disturbances and contaminants.

The transition zone is critical for a future Delta that can support terrestrial wildlife. Restoring gentle habitat transitions along a natural elevation gradient now will facilitate marsh transgression in the future as sea level rises.[18] The greatest marsh restoration opportunities are located along the periphery of the Delta because these areas are less subsided.

Contemporary photographs of terrestrial habitat types along the Delta's edge. These habitat types also formed marsh-terrestrial transition zones where they graded into emergent wetlands.

Photo Credits (left to right and top to bottom): Steve Martarano, USFWS; Daniel Burmester, CDFW; Ruth Askevold, SFEI; Marc Hoshovsky; CDFW; Ingrid Taylar; Ruth Askevold, SFEI; Daniel Burmester, CDFW

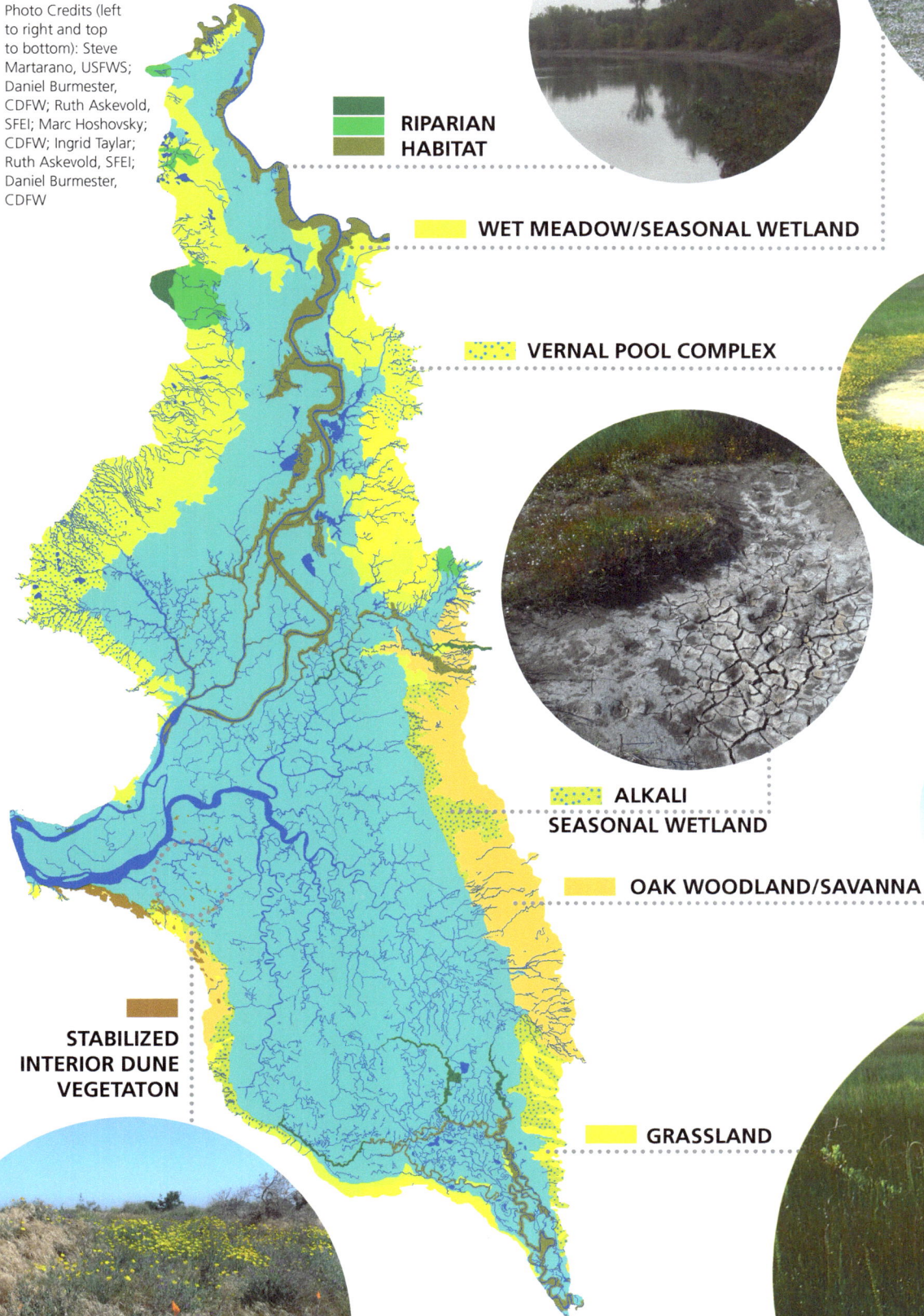

RIPARIAN HABITAT

WET MEADOW/SEASONAL WETLAND

VERNAL POOL COMPLEX

ALKALI SEASONAL WETLAND

OAK WOODLAND/SAVANNA

STABILIZED INTERIOR DUNE VEGETATON

GRASSLAND

The historical marsh-terrestrial transition zone was continuous and gradual

Today's marsh-terrestrial transition zones are fragmented

The transition zone between marsh and terrestrial habitats supported many wildlife species and ecological functions. Animals, organic matter, sediment, and water moved across this wide, complex, and heterogeneous area that supported a broad moisture gradient. Continuous transition zones bordered the Delta periphery and major riparian corridors. Most transition zones were wide and gradual, yet some were short and steep. This continuity and variability allowed diverse terrestrial wildlife to access wetland habitat, and was critical for the movement and dispersal of transition-zone obligates. The transition zone may have been particularly important to the endemic giant garter snake, which used aquatic habitats dominated by emergent vegetation from early spring to mid-fall, and drier, higher-elevation habitats during winter dormancy. Foraging birds and bats may have used seasonal wetlands at different times of the year depending on inundation and food production. In the modern Delta, the terrestrial edge is fragmented and narrow, providing less foraging access, cover, and movement corridors.

HISTORICAL

Giant
garter snake

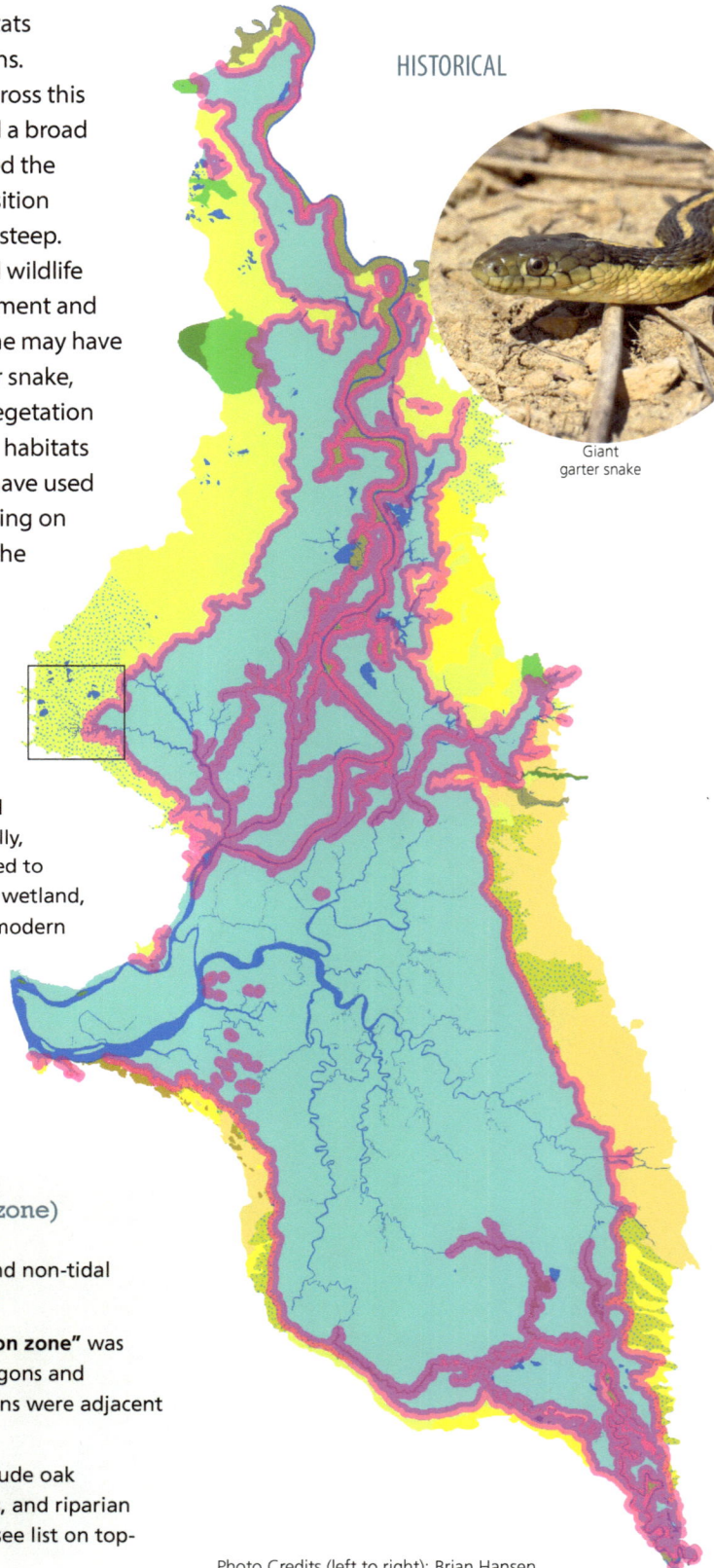

Marsh-terrestrial transition zones in the historical (right) and modern (far right) Delta, represented by pink lines. Historically, much of the marsh gradually transitioned to seasonal wetland, vernal pool, alkali wetland, or riparian forest. In contrast, the modern transition zone is discontinuous and rapidly shifts to mostly grassland. Modern grasslands are heavily altered habitats, and modern transition zones are often steep levees.

Marsh-terrestrial
transition zone

Methods: Marsh-terrestrial transition zone (T-zone)

length of t-zone

"marsh" includes both tidal and non-tidal freshwater emergent wetland

the **"marsh-terrestrial transition zone"** was mapped wherever marsh polygons and terrestrial habitat type polygons were adjacent to one another

"terrestrial habitat types" include oak woodlands, seasonal wetlands, and riparian habitat types, among others (see list on top-right of facing page)

Photo Credits (left to right): Brian Hansen, USFWS; Daniel Burmester, CDFW

MODERN

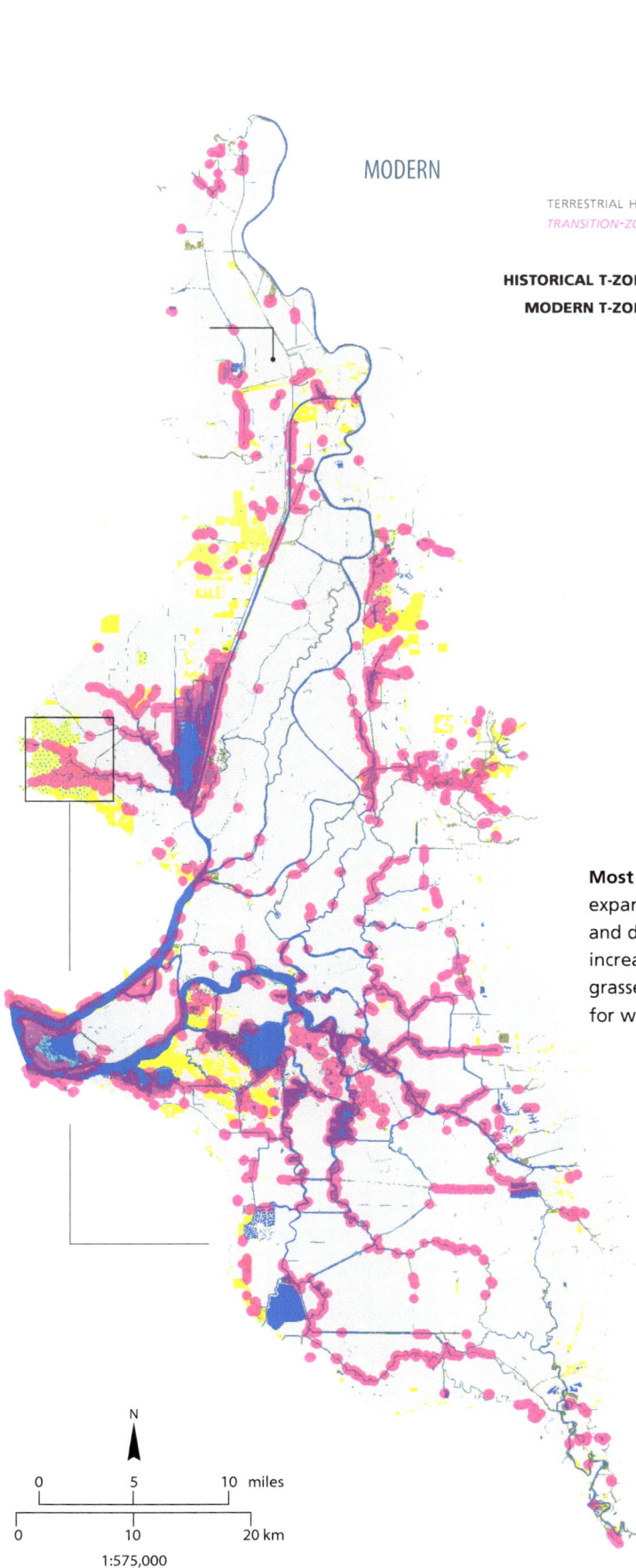

TERRESTRIAL HABITAT TYPES →	WILLOW RIPARIAN SCRUB OR SHRUB	GRASSLAND	WILLOW THICKET	VERNAL POOL COMPLEX	STABILIZED INTERIOR DUNE VEGETATION	OAK WOODLAND & OAK SAVANNA	ALKALI SEASONAL WETLAND COMPLEX	WET MEADOW & SEASONAL WETLAND	VALLEY FOOTHILL RIPARIAN
TRANSITION-ZONE (T-ZONE) →									
MARSH →									
HISTORICAL T-ZONE LENGTH (km)	108	29	17	22	39	45	81	306	607
MODERN T-ZONE LENGTH (km)	370	103	59	4	0	0	19	30	116

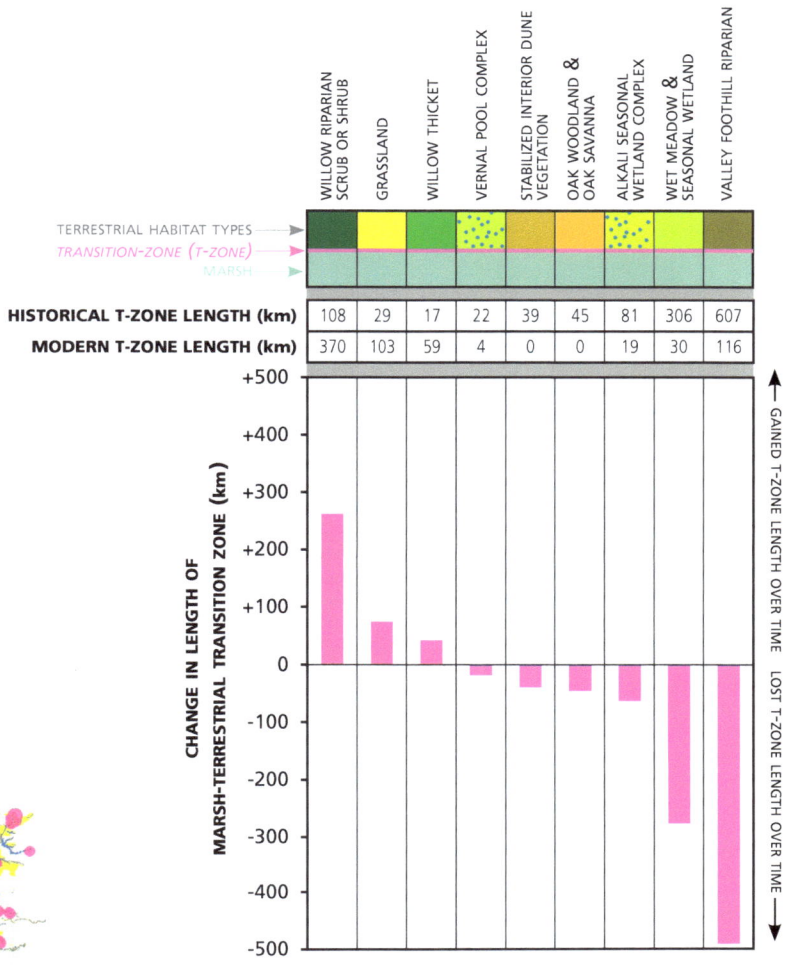

Most transition zone types are greatly reduced. A few types have expanded in quantity, but these tend to be relatively fragmented and disturbed. For example, although the extent of grassland has increased, modern grasslands are dominated by non-native annual grasses, which has changed the timing and availability of resources for wildlife.

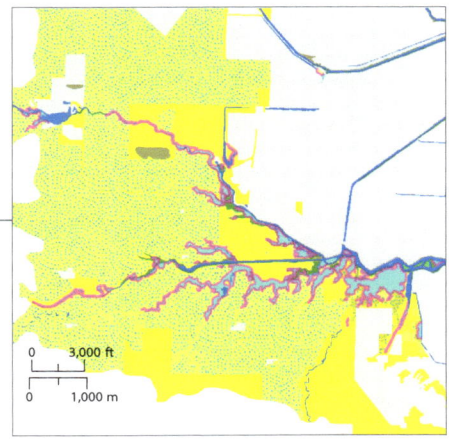

The longest continuous unleveed marsh-terrestrial transition zone left in the Delta is along Lindsey Slough (above). This area offers restoration opportunities to improve support for species using the transition zone.

Conclusion

The Delta has undergone a massive physical and biological transformation during the past two centuries. The native plant and animal species that lived and evolved in the Delta now reside in a completely different environment. With the benefit of historical research and contemporary ecological knowledge, we can infer how the pre-development Delta supported native wildlife, and identify the missing functions in today's landscape.

Most fundamentally, the historical Delta was a vast wetland complex composed of an array of habitat types, primarily freshwater marsh, defined by varying cycles of inundation. Differential patterns of flooding, from both rivers and tides, created and maintained tule marshes, lakes, seasonal wetlands, willow thickets, and riparian forests. The disconnection of natural flooding processes due to the construction of levees has profoundly altered the Delta landscape, reducing the natural resilience of the Delta's landforms and wildlife populations. The excavation of channels and building of levees created a dichotomous landscape of dry land and open water where once existed much more variable and dynamic wetlands.

Severe declines in Delta wildlife and likely future impacts from climate change and other drivers motivate a desire to restore a resilient landscape with improved wildlife support functions. Yet the major physical changes to the system, as well as the impacts from invasive species, water diversions, and other stressors, make it difficult to envision how Delta ecosystems could work successfully in the future. The native ecosystems of the Delta are altered and reduced, with few functional examples to learn from. Today's novel Delta ecosystems illustrate stressors but provide few attributes to emulate. The way forward is to design functional landscapes that can take advantage of native geomorphic templates and restorable physical and biological processes to shift the current novel Delta ecosystems toward greater wildlife support functions.

The landscape metrics presented here offer a new set of tools to analyze, design, and evaluate Delta restoration scenarios and outcomes. In the next steps of the Delta Landscapes project, the metrics and other information about past, present, and projected future conditions will be used to develop conceptual restoration visions for the Delta.

For more information, please visit: www.sfei.org/projects/delta-landscapes-project.

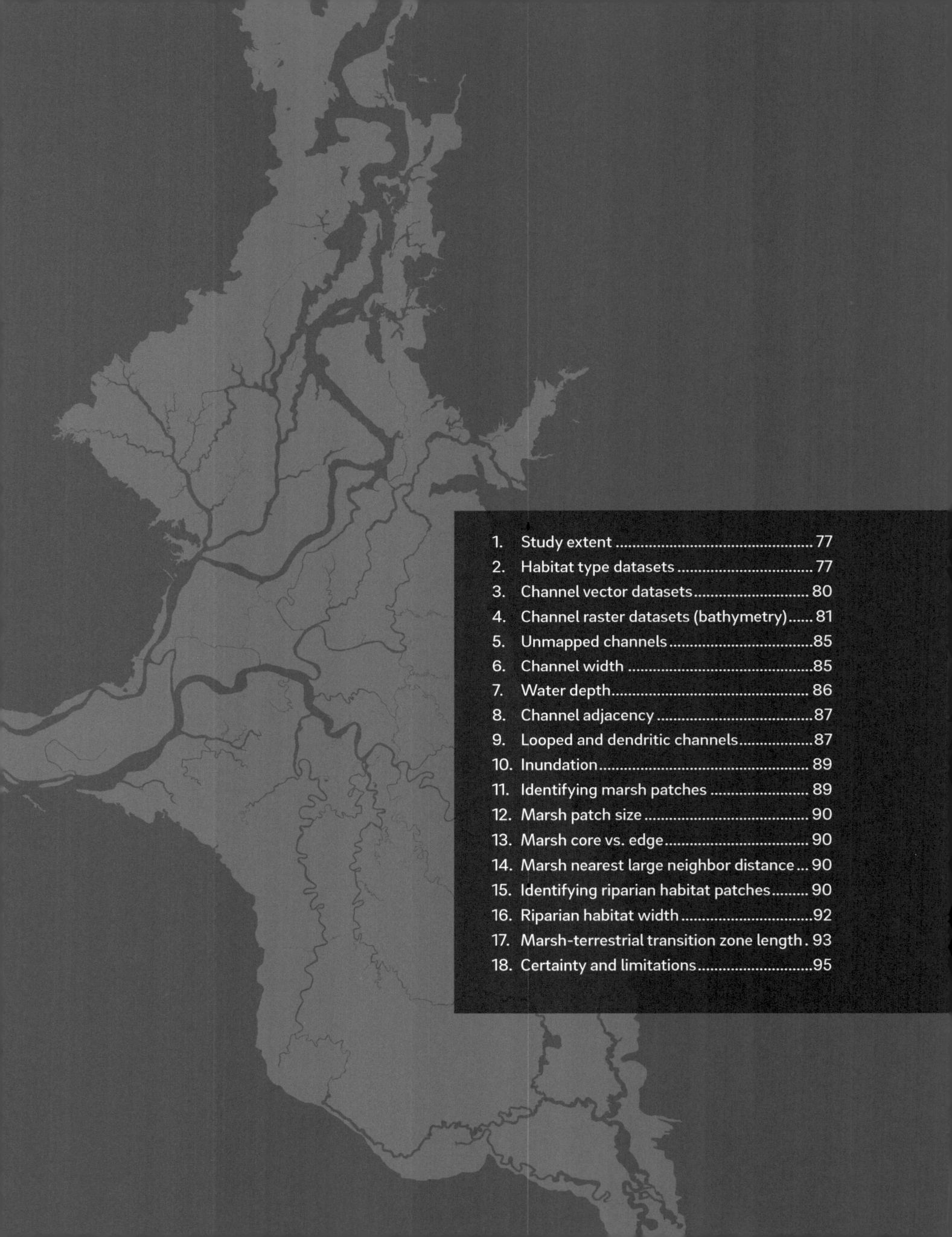

Appendix A: Methods

1. STUDY EXTENT

Our study extent is defined by the area mapped in the SFEI-ASC *Sacramento-San Joaquin Delta Historical Ecology Investigation*.[1] As detailed in that report, this area was selected to include the full extent of the Delta's historical tidal wetlands, adjacent non-tidal freshwater wetlands, and upland transitional areas. The study area was generally defined as "the contiguous lands lying below 25 feet (7.6 m) in elevation." This differs from the extent of the legal Delta and "encompasses an area of about 800,000 acres, including parts of Sacramento, Yolo, Solano, Contra Costa, and San Joaquin counties. The boundary was defined using the National Elevation Dataset (NED) 10m-Resolution (⅓-Arc-Second) Digital Elevation Model (DEM)." The report authors "used GIS tools to generalize the boundary and removed upland (fluvial) channels less than 650 feet (200 m) wide." To avoid holes in the study area, the authors included small hillocks within the outer boundary and also included areas within the sinks of Putah and Cache creeks that were above the 25 foot (7.6 m) contour.

As in the *Sacramento-San Joaquin Delta Historical Ecology Investigation*, the western boundary of this study "was established at the west end of Sherman Island in order to match the historical ecology mapping previously completed for the Bay Area EcoAtlas and Baylands Ecosystem Habitat Goals Project (Goals Project 1999)." Upstream, "the study area falls at hydrogeomorphically logical locations. On the west side of the Sacramento River, the study area extends northward in the Yolo Basin to Knights Landing Ridge, also near where the Feather River enters the Sacramento River." Not included in this or the *Delta Historical Ecology* study was "the American Basin on the east side of the Sacramento River between the American and Feather rivers as it was completely non-tidal and extended well above the 25 foot (7.6 m) contour." The southern extent of the study area was defined as the confluence of the San Joaquin and Stanislaus rivers.[2]

2. HABITAT TYPE DATASETS

2.1 Sources for the historical Delta

GIS data depicting historical Delta habitat types were obtained from SFEI-ASC's *Sacramento-San Joaquin Delta Historical Ecology Investigation* (Table 1).[3] The dataset classifies the historical Delta into 17 habitat types, the majority of which are based on modern clas-

Table 1. Sources for historical and modern habitat type datasets.

Title	Citation	Minimum mapping unit	Minimum width	Incorporated area (ha)	Study extent coverage
Historical					
Sacramento-San Joaquin Delta Historical Ecology Investigation (*'SFEI 2012 Delta HE'*)	Whipple et al. 2012	5 ha	15 m (channels only—narrower channels digitized as lines)	316,426	100%
Modern					
Vegetation and land use classification and map of the Sacramento-San Joaquin River Delta (*'CDFG 2007 Delta Vegetation'*)	Hickson & Keeler-Wolf 2007	0.4 ha (water) 0.8 ha (vegetation)	10 m	253,457	80%
Central Valley Riparian Mapping Project (*'CDWR 2012 CVRMP'*)	GIC 2012	0.4 hectares	≥10 m	60,761	19%
Natural Communities Mapping of the Cache Slough Complex vicinity from combined data sources (*'WWR 2013 CSCCA Natural Communities'*)	WWR 2013	varies	varies	725	<1%
Bay Delta Conservation Plan Natural Communities Mapping (*'CDWR 2013 BDCP Natural Communities'*)	DWR 2013	varies	varies	65	<1%
San Francisco Estuary Institute supplemental mapping (*'SFEI 2013 supplemental mapping'*)	n/a	varies	varies	1,381	<1%

sification systems (Table 2, at end of Appendix A). Readers should refer to that report for detailed methods on defining and mapping each habitat type.

2.2 Sources for the modern Delta

Since no recent effort to map modern natural communities in the Delta covers the entire study extent, modern habitat data were compiled from multiple sources (Table 1) and then crosswalked, when possible, to the historical habitat types used by Whipple et al. (2012) (Table 3, at end of Appendix A; see Section 2.3 for information on the crosswalk utilized in this study). Additional habitat types were incorporated into the modern classification system when analogues to historical classifications were unavailable (e.g., 'Managed wetland,' 'Agriculture/Non-native/Ruderal,' and 'Urban/Barren').

The Vegetation Classification and Mapping Program's (VegCAMP) 2007 Sacramento-San Joaquin River Delta dataset ('CDFG 2007 Delta Vegetation')[4] served as the primary component of our modern habitat type layer. This mapping effort utilized true color 1-foot resolution aerial photography from the spring of 2002 (and from the summer of 2005 in some marginal areas) to classify 129 fine-scale to mid-scale vegetation mapping units within the extent of the legal Delta. Although the dataset is derived from imagery that is now more than a decade old, it is still the most comprehensive (with respect to extent and resolution of vegetation mapping units) available for the Delta. Eighty percent of our Modern Habitat Type layer was derived from this source.

Since our dataset extended beyond the boundaries of the legal Delta, the 'CDFG 2007 Delta Vegetation' dataset was supplemented with VegCAMP's 2012 Central Valley Riparian Mapping Project Group Level dataset ('CDWR 2012 CVRMP').[5] This mapping effort utilized 2009 National Agricultural Inventory Program (NAIP) aerial imagery, from which polygons were hand-digitized. Nineteen percent of our Modern Habitat Type layer was derived from this source.

When combined, the 'CDFG 2007 Delta Vegetation' and 'CDWR 2012 CVRMP' datasets provided coverage for more than 99% of our study extent. Remaining data gaps were filled with a combination of sources, including an unpublished natural communities dataset developed for the Cache Slough Complex Conservation Assessment (itself a combination of sources compiled by Wetlands and Water Resources, Inc.; '2013 CSCCA Natural Communities')[6] and a natural communities dataset developed for the Bay Delta Conservation Plan ('CDWR 2013 BDCP Natural Communities').[7] Polygons for the remaining areas without coverage were hand-digitized and classified by SFEI staff using Bing aerial photographs accessed in 2013 ('SFEI 2013 supplemental mapping').

A map displaying where each dataset was used to develop the modern habitat type layer can be found on page 15.

2.3 Historical-modern crosswalk

To compare the historical and contemporary landscape, we were required to crosswalk the detailed modern classifications (from each of the modern datasets listed above and in Table 1) to the habitat types utilized in the historical habitat types layer. The crosswalk from 'CDFG 2007 Delta Vegetation' mapping units to the historical habitat types was developed for the *Sacramento-San Joaquin Delta Historical Ecology Investigation*[8] with the help of local experts (Table 3, at end of Appendix A).[9] Since the historical habitat types were based on modern classification systems, the crosswalking process was generally straightforward. However, several map units classified in the 2007 mapping were challenging to associate with a historical classification. It was determined that "Distichlis spicata- Annual Grasses," for example, should be placed in the "Wet meadow or seasonal wetland" category instead of the "Alkali seasonal wetland complex" category, as the area where it was extensively mapped (in the Yolo Bypass) is characterized by conditions more similar to the wet meadow or seasonal wetland type used for mapping the historical Delta.[10] Willow-dominated communities also posed challenges. The crosswalk attempted to group the modern alliances based on the historical habitat classification of whether the willows were part of a backwater swamp community (willow thicket), the dominant species along channel banks (willow riparian forest, scrub, or shrub), or were part of a forest with oaks (valley foothill riparian forest).

Since the fine-scale (mostly Alliance level) classifications of 'CDWR 2012 CVRMP' were derived from the 'CDFG 2007 Delta Veg' map, our crosswalk developed for the 2007 Delta layer was also applicable to the 2012 Central Valley layer. The medium-scale (mostly Group level) classifications of 'CDWR 2012 CVRMP,' however, had no existing crosswalk. The crosswalk for this dataset (presented in Table 3, at end of Appendix A) was developed by SFEI staff from group characteristic vegetation descriptions[11] and with input from local experts.[12]

'CDWR 2013 BDCP Natural Communities' and '2013 CSCCA Natural Communities' layers utilized the Multi-Species Conservation Strategy NCCP Habitat Types classifications,[13] which had already been related to the historical classification types (Table 2, at end of Appendix A) and were therefore simple to crosswalk (Table 3, at end of Appendix A).

For some purposes, the classifications established by the crosswalks were modified based on additional data and criteria (see Section 2.4).

2.4 Deviations from established crosswalk

2.4.1 Willow-marsh complex

Many polygons in the modern dataset classified as 'Freshwater emergent wetland' are ringed by a strip of vegetation classified as 'Willow thicket.' Conversations with California Department of Fish and Wildlife scientists and further examination of the underlying vegetation types crosswalked to 'Willow thicket' indicated that a significant percentage of polygons contained some freshwater emergent wetland species and thus might be considered part of a larger willow-marsh complex.[14] To capture this unique landscape feature also reported historically in the Central Delta,[15] we reclassified 'Willow thicket' polygons that contained freshwater emergent wetland species (and thus indicated a lower, wetter environment) as 'Willow-marsh complex.' We also selected contiguous 'Freshwater emergent wetland' polygons that intersected the new 'Willow-marsh complex' polygons and reclassified these as 'Willow-marsh complex.' Most of the modern Delta's in-channel marsh islands are classified as 'Willow-marsh complex.' For many metrics, 'Freshwater emergent wetland' and 'Willow-marsh complex' are lumped during analysis. This reclassification was particularly important for metrics addressing the marsh-water edge (since freshwater emergent wetlands ringed by a thin strip of willow thicket would not have any such edge). A list of the map units that composed the original 'Willow thicket' habitat type and an account of which units were reclassified as 'Willow-marsh complex' can be found in Table 4.

2.4.2 Managed wetlands

For the modern habitat type layer we sought to distinguish managed wetlands (characterized by novel forms and managed hydrographs, often separated from direct tidal action by tide gates and weirs, and commonly constructed to support waterfowl) from other wetland areas. Managed wetlands were identified with BDCP's Natural Communities dataset (2009-2013). Polygons with the 'SAIC_Type' of 'Managed wetland' were extracted from the BDCP layer and incorporated into our modern habitat map with ArcGIS's 'Union' tool. Since both datasets were compiled, in large part, from CDFW's Delta Vegetation dataset,[16] alignment between the two datasets was quite high. Additional managed wetlands were identified by SFEI staff from modern aerial images.

2.4.3 Riparian connectivity

Not all polygons in the modern dataset classified as riparian vegetation types ('Valley foothill riparian' and 'Willow riparian scrub/shrub') are hydrologically connected to an adjoining channel. To distinguish between functionally riparian vegetation and hydrologically disconnected riparian-type vegetation, we created two new habitat subtypes. The 'Valley foothill alliance' and 'Willow scrub/shrub alliance' classifications represent hydrologically disconnected polygons originally classified as 'Valley foothill riparian' or 'Willow riparian scrub/shrub,' respectively. A polygon was considered hydrologically connected if it shared an edge with a polygon classified as 'Water.' Riparian polygons that were connected to water through other riparian polygons (of either type), polygons classified as 'Freshwater emergent wetland,' and/or polygons classified as 'Willow-marsh complex' were also considered hydrologically connected. This analysis was meant only to approximate hydrologic connectivity at a coarse level—it does not, for example, distinguish between standing water and creeks, nor does it consider topography or flood frequency. Not all analyses use the split classifications—for some (where vegetation type and structure is more important than hydrology), the original, more general classifications are used. See Section 16.2 for a map of hydrologically connected and disconnected riparian habitat.

2.4.4 Vernal pool complex

It became apparent that much of the 'CDFG 2007 Delta Veg' map units initially crosswalked to 'Grassland' were likely better represented as 'Vernal pool complex.' The same issue was addressed by the BDCP Natural Community Mapping effort (CDWR 2013, Appendix 2.B), which assembled a Vernal Pool Review Team to classify and map vernal pool complexes within the BDCP Plan Area. The BDCP classifications were informed by a number of datasets, including the Soil Survey Geographic Database (SSURGO), the BDCP composite vegetation GIS layer, Google Earth aerial imagery, 2007

Table 4. Reclassifying *Willow thicket* for modern habitat type layer.

Map units originally classified as Willow thicket	Reclassification
Buttonbush (*Cephalanthus occidentalis*)	Willow thicket
California Dogwood (*Cornus sericea*)	Willow thicket
California Hair-grass (*Deschampsia caespitosa*)	Willow thicket
Cornus sericea - *Salix exigua*	Willow thicket
Cornus sericea - *Salix lasiolepis* / (*Phragmites australis*)	Willow-marsh complex
Salix lasiolepis - (*Cornus sericea*) / *Scirpus* spp.- (*Phragmites australis* - *Typha* spp.) complex unit	Willow-marsh complex
Shining Willow (*Salix lucida*)	Willow thicket

LiDAR elevation data, California Natural Diversity Database (CND-DB) records, existing management and habitat conservation plans, and vernal pool expert knowledge.[17]

In light of this focused effort, we replaced polygons crosswalked as 'Grassland' in our preliminary modern habitat types layer with polygons identified as 'Vernal pool complex' by the BDCP mapping effort whenever the two overlapped. Specifically, we used the 'Union' tool in ArcGIS to replace polygons from 'CDFG 2007 Delta Vegetation' and 'CDWR 2012 CVRMP' crosswalked to 'Grassland' with those polygons from '2013 CSCCA Natural Communities' that had an "SAIC_Type" of 'Vernal pool complex' (a classification derived from the 'CDWR 2013 BDCP Natural Communities' layer). Where polygons from the two datasets overlapped, the habitat type was changed to 'Vernal pool complex' (otherwise the habitat type remained 'Grassland').

2.4.5 Swale form

'CDWR 2012 CVRMP' polygons with an "NVCS_NAME" of "California annual forb/grass vegetation" were initially crosswalked to 'Grassland.' However, when these polygons exhibited the natural swale form common to the edge of alluvial fans between ridges on the eastern and western edges of the Delta, the 'Grassland' classification was changed to 'Wet meadow/Seasonal wetland.' This reclassification better captures the hydrology and landscape position of these features, which are natural, seasonally wetted low spots on the landscape that generally offer potential for upland transgression of marshes with sea level rise. Additionally, the reclassification provides greater alignment with the habitat type assigned to these landforms by the finer-resolution 'CDFG 2007 Delta Vegetation' mapping and crosswalk.

2.5 Non-native and invasive species

We sought to map areas in the modern Delta where invasive or non-native plant species are dominant or co-dominant with native vegetation and to quantify the percent area dominated by non-native/invasive vegetation by habitat type. Individual habitat type polygons were marked as dominated by non-native/invasive vegetation if their vegetation mapping unit (generally associated with alliance- and association-level classifications) featured a non-native species (as defined by CalFlora) or invasive species (as defined by Cal-IPC). Where alliance/association-level classifications were unavailable, the non-native/invasive designation was determined based on Group-level classifications and best professional judgement. Table 5 (at end of Appendix A) lists the mapping units of the modern habitat type layer and whether or not each was classified as dominated by non-native/invasive vegetation. For the purposes of the map, we also classified areas

with a habitat type of either "Agriculture/Non-native/Ruderal" or "Urban/Barren" as non-native, regardless of the more specific mapping unit classification.

3. CHANNEL VECTOR DATASETS

GIS layers of Delta hydrography were required to develop the project's suite of channel-related metrics. Since both forms of data were needed for our analyses, we obtained or generated polygon and polyline datasets of channel hydrography in the historical and modern Delta (unlike polygons, polylines are one-dimensional features with no width or area in the GIS). From these geodatasets, we developed metrics of channel length, width, adjacency, density, and sinuosity (the latter two are not presented in this report). We also classified channel reaches as either "dendritic" or "looped" (see Section 9 for definitions of these terms). Maps of these datasets can be found on page 15.

3.1 Sources for the historical Delta

Historical Delta channel polygons were obtained from the SFEI-ASC *Sacramento-San Joaquin Delta Historical Ecology Investigation*'s historical habitats layer.[18] The SFEI-ASC study generated polygons for channels at least 15 m in width and 50 m in length and incorporated these features into the map of historical Delta habitat types. For use in developing channel-related metrics, polygons classified as 'fluvial low order channel,' 'fluvial mainstem channel,' 'tidal low order channel,' or 'tidal mainstem channel' were extracted from the habitat type layer and clipped to the study extent.

Historical Delta channel polylines were obtained from the SFEI-ASC *Sacramento-San Joaquin Delta Historical Ecology Investigation*'s historical creeks layer.[19] This line layer contained all channel features longer than 50 m, regardless of their width. For channels also digitized as polygons, the polyline layer represents an approximate channel centerline. The layer was edited to ensure that channel polylines associated with polygons fell completely within the polygon boundaries.

3.2 Sources for the modern Delta

Modern Delta channel polygons were obtained from a 2013 version of the National Hydrography Dataset (NHD),[20] clipped to the project study extent, and selected by feature type to isolate features classified as 'StreamRiver' or 'CanalDitch.'

Modern Delta channel polylines were generated from the NHD polygons (described above) with a custom centerline generation tool. To best match the historical dataset, channel polylines were only generated around islands greater than 25 ha in size. If an island did not meet this size requirement, it was considered an "in-channel" island (an island located within a single channel as opposed to an island bounded by multiple, separate channels), and dissolved with the channel polygon for the purposes of centerline generation. To generate the channel polyline layer from the NHD polygons, the centerline tool converted NHD polygons to outlines, added additional vertices to these lines every 10 meters, created points from the vertices, and calculated Theissan polygons from the points. These Theissan polygons were converted to outlines, which were then clipped to the NHD polygons and split at vertices. The tool then removed segments less than 100 m with dangling ends, merged and exploded all lines, and then deleted all lines that were not connected on both ends and consisted of only 2 vertices (leaving only the polygon centerlines). To eliminate channels associated with small man-made harbors, we manually removed resulting reaches that were both less than 500 m in length and deemed unnatural. After evaluating the resulting layer, we digitized additional channel polygons and polylines that were not in the original NHD dataset, including the tidal channel networks of Sherman Island, Mandeville Tip, the Liberty Island Conservation Bank, the Cosumnes Floodplain Mitigation Bank, and along the Yolo Bypass Toe Drain. Like with the historical dataset, there was no minimum width employed for digitizing a channel line.

4. CHANNEL RASTER DATASETS (BATHYMETRY)

To develop metrics involving channel depth, we obtained or generated rasters of channel bathymetry for both the historical and modern time periods. Using this elevation data, we developed approximations of water depth at specific tidal datums (see Section 10). As described below, the historical Delta DEM was developed for a separate project and was constructed at a 2 m resolution (to capture the smallest channels). The modern Delta DEM, a California Department of Water Resources product, is an integrated 2 m and 10 m resolution raster. Both DEMs were clipped to include only the mutually mapped areas.

4.1 Sources for the historical Delta bathymetry

It is the goal of a separate, ongoing project to characterize the hydrodynamics of the San Francisco Bay-Delta Estuary under more natural conditions (those prior to major modification of Bay-Delta geometry and hydrology beginning in the mid-19th century) through the development and use of a 3D hydrodynamic model. One critical task of this larger project is the creation of a bathymetric-topographic digital elevation model (DEM) of the early 1800s historical Delta. The development of this raster is a collaborative effort between researchers and technicians at the San Francisco Estuary Institute (SFEI), the Center for Watershed Sciences (CWS) at the University of

California, Davis, and Resource Management Associates, Inc. (RMA), funded by CWS and the Metropolitan Water District. Please see Table 6 for a list of individuals who have contributed to the development of the historical Delta DEM used in this report.

A manuscript with the methods used to develop the historical Delta DEM is currently in preparation for publication.[21] Here, we provide a simplified overview of the methods used to develop the historical bathymetry raster utilized in this report. Greater details on the development of the dataset (including the topographic component, which is not used or discussed in this report) will be available in the near future. Since the project is ongoing, the historical DEM used in this report constitutes an interim product and is subject to future modification.

This report utilizes version 3.1 of the Historical Delta Topographic-Bathymetric DEM, an interim product released internally in July 2014. To create this DEM, the project team integrated 2D historical Delta channel planform and land cover data from previous mapping efforts (Whipple et al. 2012) with elevation data from numerous historical sources. Raw historical bathymetric data were obtained primarily from mid-19th century sources, including U.S. Coast Survey (USCS) hydrographic sheets and early river surveys. Different areas and components of the Delta had to be addressed separately, given data availability. Three general sets of methods were used and combined to develop the DEM bathymetry (Figure 1).

Table 6. Individuals who have contributed to the development of the historical Delta digital elevation model (DEM) used in this report (alphabetical by institution). This work is being conducted as part of a separate, ongoing project (funded by CWS and the Metropolitan Water District).

Contributors
Center for Watershed Sciences (CWS)- University of California, Davis
Andrew Bell
William Fleenor
Mui Lay
Amber Manfree
Alison Whipple
San Francisco Estuary Institute (SFEI)
Julie Beagle
Robin Grossinger
Samuel Safran
Resource Management Associates, Inc. (RMA)
Stephen Andrews
John DeGeorge

HISTORICAL

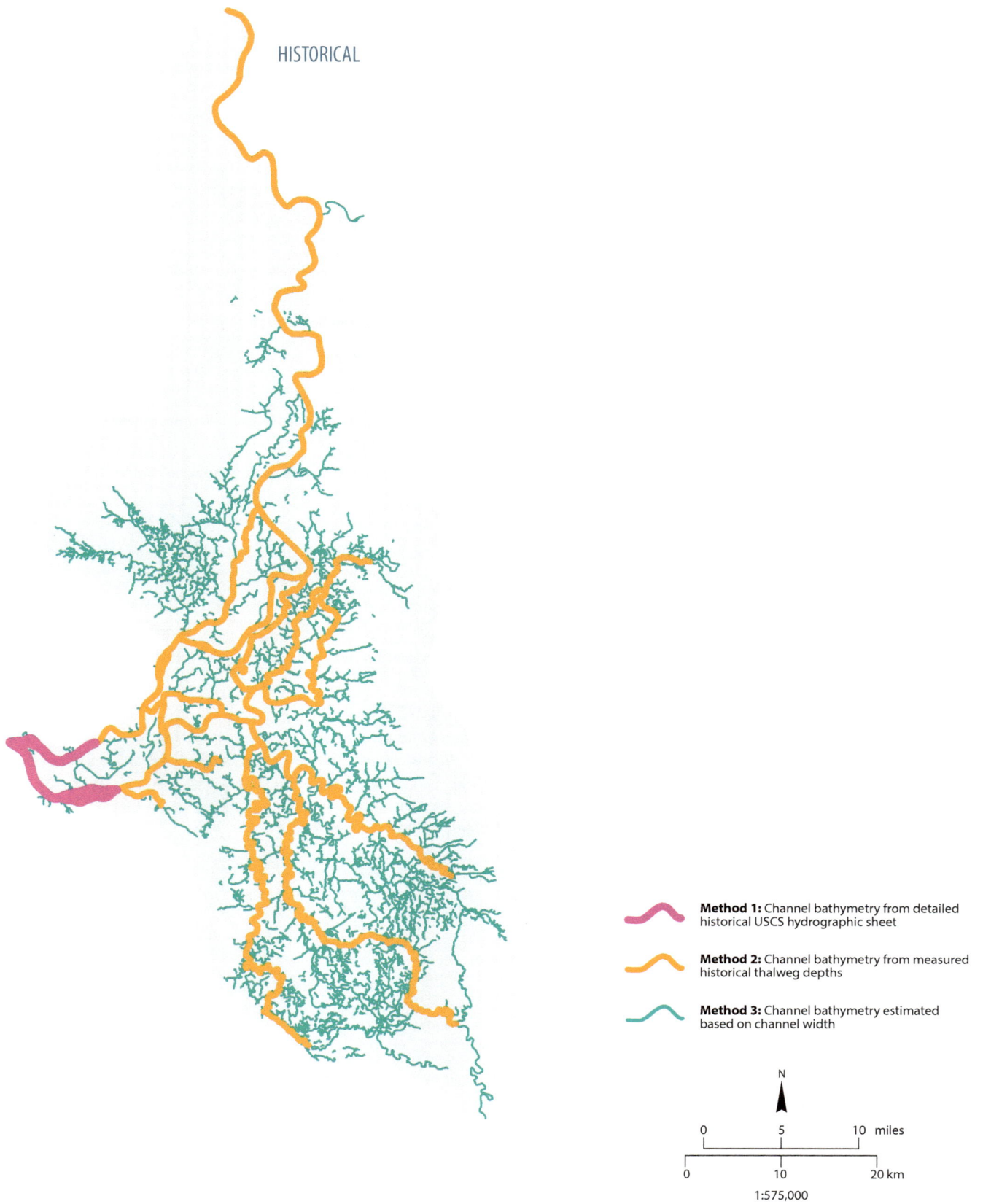

Method 1: Channel bathymetry from detailed historical USCS hydrographic sheet

Method 2: Channel bathymetry from measured historical thalweg depths

Method 3: Channel bathymetry estimated based on channel width

N

| 0 | 5 | 10 miles |

| 0 | 10 | 20 km |

1:575,000

Figure 1. Three methods used to develop historical Delta bathymetry. Methodology varied based on data availability. See section 4.1 of this chapter for a more detailed description of each method.

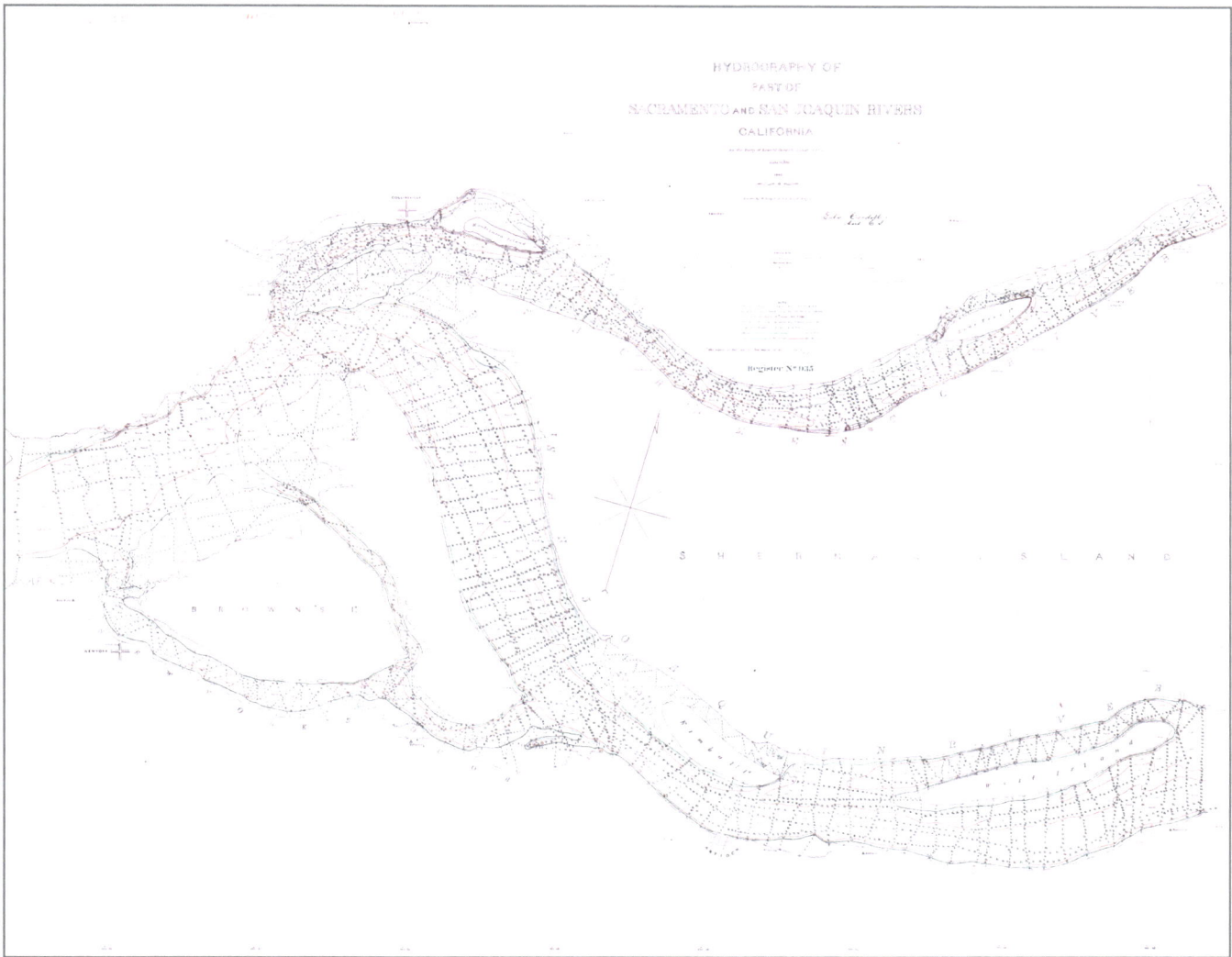

Figure 2. Cordell 1867 (United States Coast Survey), "Hydrography of part of Sacramento and San Joaquin Rivers California." This map is a hydrographic sheet ("H-Sheet") with historical bathymetry of the Delta mouth. The bathymetry seen here was digitized directly from a georeferenced version of the map.

4.1.1 Method 1: Bathymetry of the Delta mouth

A detailed 1867 U.S. Coast Survey hydrographic sheet with historical bathymetry was available downstream of Sherman Island (Figure 2).[22] The project team digitized 4,809 soundings and three bathymetric contour lines (6, 12, and 18 ft) directly from a georeferenced version of this map (which indicated depth at mean lower low water [MLLW]). After converting the digitized soundings to a modern fixed datum (NAVD88, see Section 4.1.4), the points were used directly as TIN inputs to generate continuous DEM bathymetry.

4.1.2 Method 2: Bathymetry of channels with measured historical data

The U.S. Coast Survey produced detailed 19th century bathymetric maps for the San Francisco Bay Estuary only as far upstream as Sherman Island. Bathymetry upstream of this location was derived from three historical river surveys (Ringgold 1850a, Ringgold 1850b, and Gibbes 1850), each conducted before the extensive mid- to late 19th century hydraulic mining in the Sierra Nevada foothills that altered bed elevations in the Delta. Critical locations were substituted with soundings from maps created by the California Debris Commission between 1908 and 1913.[23] Unlike the USCS (1867) hydrographic sheet, the historical river surveys generally only indicated the depth of the deepest part of the channel (the channel "thalweg"). Soundings were generally taken or adjusted by the surveyors to low water conditions. In total, the project team georeferenced 1,484 historical soundings indicating mean lower low water thalweg depth.[24] We snapped georeferenced points to a historical thalweg polyline and interpolated thalweg depths between these points using a spline function. We assumed a parabolic channel shape to generate bathymetry on either side of the thalweg.[25] Channels with bathymetry derived from this method are shown in Figure 1.

4.1.3 Method 3: Bathymetry of channels without measured historical data

Measured historical soundings were not available for much of the study extent. Because of this, we sought to determine historical channel depths by generating a regression relating channel depth to channel width. The relationship between these two variables was determined with available measured historical thalweg depths (described above). Historical channel widths (see Section 6) were spatially joined to historical soundings to create a dataset of historical channel widths with associated MLLW thalweg depths. Measured historical widths and depths were plotted against one another and fitted with a power function (Figure 3). A power function was selected because of known power relationships between width and depth in fluvial systems and because it avoided generating negative depths at smaller channel widths. While not perfect, this method was selected after extensive conversations with experts on tidal marsh morphology, and appears to provide reasonable estimates of channel depth given the available information. The function took the following form and was used to extrapolate depths for all channels:

Let y = channel depth at MLLW

Let x = channel width

$y = 0.8516x^{0.4111}$

$R^2 = 0.34$

Small historical channels (with widths <15 m) were originally digitized as polylines and thus did not have a precisely known width for use in the regression. We assigned these channels a width of 7 m (approximately half the minimum mapping unit for digitizing channels polygons) when extrapolating depths using the width-depth regression.

4.1.4 Converting a historical tidal datum (MLLW) to modern fixed datum (NAVD88)

Historical sources for bathymetry were created well before the development of a standardized vertical datum (such as the Sea Level Datum of 1929) and were simply referenced to a low water surface. To use the historical Delta DEM in hydrodynamic models, the project team converted the historical MLLW data to a modern fixed datum (NAVD88). The method utilized in version 3.1 of the historical Delta DEM entailed two primary steps: (1) converting historical mean lower low water (MLLW) depths to historical local mean sea level (MSL) depth by adding tidal amplitude (or one-half the tidal range) to MLLW depth and (2) obtaining historical bed elevations (in NAVD88) by subtracting MSL depth from MSL elevation (in NAVD88). To implement these steps, the project team was required to determine two variables: (1) historical Delta tidal range and (2) historical Delta mean sea level elevation, both of which vary spatially. Rasters quantifying these variables across space were developed to convert historical depths to NAVD88.

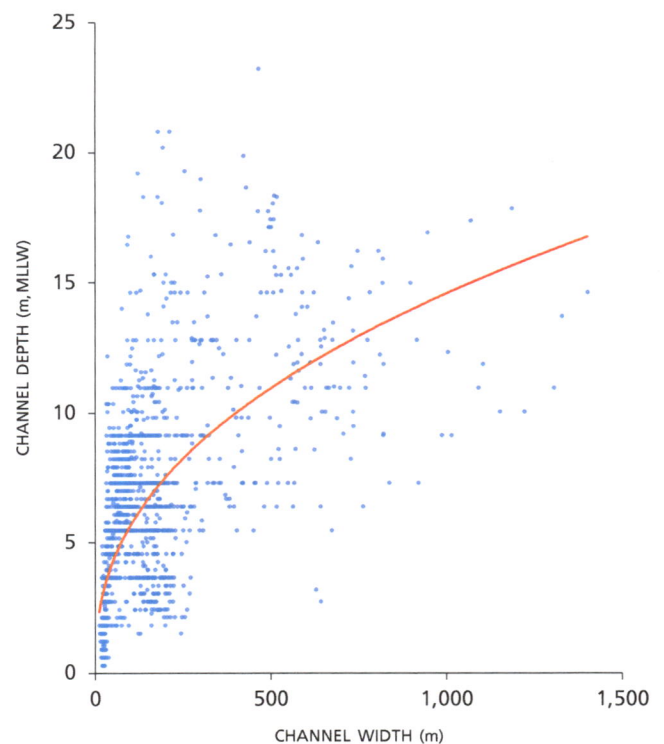

Figure 3. Scatter plot of historical channel depth vs. historical channel width. Each data point represents one historical sounding (adjusted to MLLW and representative of the thalweg depth) plotted against the width of the historical channel at the sounding's location (as derived from the historical channel polygon layer; N = 1,484). Data points have been fitted with a power function (red line) with the above equation.

The historical tidal range surface utilized in version 3.1 of the historical Delta DEM was developed by interpolating between georeferenced historical textual data using a natural neighbors method and the 'Create TIN' tool in ArcGIS. Additional points with a tidal range of 0 were created at the boundary between tidal and nontidal channel reaches as mapped by Whipple et al. (2012). Where records were too far apart for the TIN to successfully/realistically interpolate between them, best professional judgment was used to add values between known points. In total, 75 georeferenced points of historical tidal range were used to generate the historical tidal range surface.

The historical MSL surface utilized in version 3.1 of the historical Delta DEM was modeled using the RMA-2V model of the contemporary Delta, minus Delta exports and the most significant channel cuts, gates, and barriers. This simulation also subtracted estimated sea level rise (SLR) since the historical period, assuming an average rate of 0.1-0.2 cm/yr.[26]

Adjusted bathymetry was exported as a 2 m DEM and clipped to the tidal open water portions of the historical Delta, as mapped in the Delta Historical Ecology Investigation.[27]

4.2 Sources for the modern Delta bathymetry

Modern bathymetry was extracted from a continuous topographic-bathymetric DEM of the San Francisco Bay-Delta Estuary developed by California Department of Water Resources staff.[28] To facilitate comparison with the historical bathymetry raster (which was clipped to tidal open water features), we clipped the modern raster to include only cells with subtidal elevations. Subsided islands surrounded by levees posed a problem, since these areas have elevations well below sea level but are not actually aquatic habitat. Because the modern DEM features numerical orthogonal reinforcement of levees around islands,[29] we were able to exclude subtidal elevations associated with subsided islands by using the 'Magic Erase' tool in ArcGis 10.1 (ArcScan extension). We reclassified raster cells into supratidal and inter/subtidal elevations (above and below a mean higher high water elevation of 195 cm NAVD88)[30] and then used the tool to select subtidal cells directly connected to the Sacramento-San Joaquin river confluence/tidal source. Inter/subtidal areas ringed by supratidal levees were not connected and thus not selected. This process identified two subsided islands with underresolved/unenforced artificial levees in the modern DEM. We manually modified the suspect cells to enforce these levees and exclude the subsided areas within them.

5. UNMAPPED CHANNELS

It is likely that at least one class of low-order tidal channels existed in the Delta that was not represented by historical sources and was thus under-represented in the historical mapping of the Delta.[31] To match the detail and minimum mapping unit of the modern channel dataset, we sought to estimate the length of these "unmapped" historical channels in the study extent and to account for them in our analyses.

No remnant marshes with intact channel networks exist in the modern Delta from which to estimate historical channel density. General agreement exists that the channel density observed now at Sherman Island (~70 m/ha) is higher than it was historically due to the relatively young age of the system (until recently, Sherman Island was a depository for dredge spoils and the channel network observed today is likely overly-interconnected and under-developed as a result).[32] Length of unmapped channels was therefore estimated based on observed historical tidal channel densities in regional freshwater tidal marshes. Grossinger (1995) used USCS T-Sheets to calculate historical tidal marsh channel densities in

the upper reaches of the Napa River, where freshwater influence is dominant—the upper-two systems in Napa were found to have historical densities of 19 and 51 m/ha.[33] Collins and Grossinger (2004) also calculated a historical channel density of approximately 30 m/ha in the freshest Bay Area systems.[34] These values agree with the highest local mapped densities in the historical Delta of 30 m/ha.[35] Weighing this evidence, we established low- and high-end estimates of Delta channel density of 20 m/ha and 40 m/ha, respectively. Since these estimates are for regularly inundated tidal marshes with developed channel networks, they were only applied to areas classified as tidal freshwater emergent wetland within the area thought to experience daily tidal inundation (see Section 10).

Mapped channel density in the study extent (14.76 m/ha) was determined by dividing the length of mapped tidal channels within regularly inundated tidal freshwater emergent wetland (1,129,158 m) by the area of the regularly inundated tidal freshwater emergent wetland itself (76,506 ha). Given the mapped density, the additional unmapped channel length needed to reach our low (20 m/ha) and high (40 m/ha) density estimates was calculated to be 400,960 m and 1,931,080 m, respectively.

6. CHANNEL WIDTH

Channel width was determined by casting perpendicular transects from the channel polyline layer, trimming these transects with the channel polygon layer, and then attributing the lengths of the trimmed transects back to the channel polyline. Prior to this analysis, versions of the historical and modern polyline layers were smoothed with a maximum offset of 0.2 meters to eliminate small sharp angles in the polyline (legacies of the original digitization process). Transects were cast at 100 m intervals perpendicular to the smoothed polyline and then trimmed with the channel polygon layer. Trimmed transects were then used to segment the original channel polyline layer. Channel width was calculated for each of the resulting segments by averaging the length of transects intersecting the segment (generally one transect at each end of the segment).

Prior to trimming transects with the channel polygon layer, the channel polygons were first dissolved and then split manually at confluences to eliminate overestimations of channel width where channels converge. The overall channel width analysis was also complicated by the existence of numerous islands located within channels. For islands greater than 25 ha in size, separate channels were drawn on either side of the island and each assigned their own widths. If an island was less than 25 ha, however, it was considered an island within a single channel. When calculating channel width, we only measured the width of the water, excluding width associated with in-channel islands. For the historical channel width analysis, channels digitized only as polylines were assigned a width of 7 m (approximately half the minimum mapping unit used for digitizing channels as polygons).

7. WATER DEPTH

In Section 4 above, we described the process of developing rasters of channel bed elevations (channel bathymetry) in the historical and modern Delta. This section describes the process of using these rasters to develop approximations of water depth at a specific tidal datum.

Water depths were derived from the raster datasets of historical and modern bathymetry, which were clipped to exclude supratidal habitat (described in Section 4). Since these rasters quantify bed-elevations, we were required to establish water surface elevations to determine water depth. In the absence of comprehensive spatial datasets indicating the elevations of tidal datums to relate geodetic data to tide heights (for both the historical and modern Delta), we opted to measure depth from a single water surface elevation across the Delta. In the modern Delta, the water surface was set to 0.64 m NAVD88, a mean lower low water (MLLW) elevation calculated from various monitoring data in the Cache Slough Complex.[36] For the historical Delta, we made the simplifying assumption that the only changes to the elevation of MLLW since the historical period are from sea level rise (SLR). This assumption discounts any changes in Delta water surface elevations caused by large-scale changes like channel geometry modification, channel armoring, subsidence, or water exports). Assuming a SLR rate in the Delta of 2 mm/year during the historical period, we estimated that 0.33 m of SLR occurred between 1850 and 2013. This factor was subtracted from the contemporary elevation of MLLW at the Cache Slough Complex (used as the water surface elevation in the modern analysis) to yield a historical water surface elevation of 0.31 m. The values used to bin bed-elevations (m, NAVD88) into water-depth classes (m, MLLW) based on these water surface elevations can be found in

Table 7. The values used to bin bed-elevations (m, NAVD88) into water-depth classes (m, MLLW). The process for setting the historical and modern water surface elevations is described in Section 7.

Water depth (m, MLLW)	Bed elevation (m, NAVD88)	
	Historical	Modern
0 m (reference plane/water surface elevation)	0.31	0.64
1	-0.69	-0.36
2	-1.69	-1.36
5	-4.69	-4.36
10	-9.69	-9.36

Table 7. Water-depth classes were chosen based on input from the Landscape Interpretation Team (Chapter 2, page 6) and meaningful photic zones.[37]

7.1 Depth by area

Using the values listed in Table 7, we calculated the area of habitat in each depth-class using the 'Build Raster Attribute Table' tool in ArcGIS 10.1 and multiplying the cell count in each bed-elevation/water-depth range by cell area.

Historical perennial tidal lakes were not accounted for in the version of the historical Delta DEM utilized in this report (version 3.1). To account for these features in our analysis of historical depth, we extracted historical habitat type polygons classified as 'Tidal perennial pond/lake' and then assigned these polygons with depths obtained or derived from the available historical data. Some lakes (such as Secret Lake and Beaver Lake in the north Delta) have specific historical accounts describing their depths. When available, we used this information to assign the lakes a maximum depth, and then used buffers to generate concentric rings at each of the shallower depth classes (we assumed depth increased linearly from 0 m at the edge of the lake to the maximum depth at the center). The majority of mapped historical lakes, however, did not have lake-specific data on historical depths. For these features, we assigned inferred depths based on more general regional accounts. Historical sources, as reported by Whipple et al. (2012) suggest that many lakes in the north Delta (even large ones) were "only a few feet below the general elevations of the basins. Early travelers . . . could wade across." Considering this, we assigned most of the North Delta lakes a depth of 0–1 m. The centers of larger lakes (where more than 1,000 ft from the lake's edge) were placed in the 1-2 m depth class. This distance was relatively arbitrary, but was chosen to give the larger lakes a three-dimensional shape. In the south Delta, Whipple et al. (2012) note that historical descriptions of "knee-deep" water suggest relatively shallow features and that a map from 1850 "includes soundings of six to nine feet (1.8-2.7 m) of water in a lake." Weighing this evidence, we used a 300 m buffer to assign the centers of the larger south Delta lakes to the 1-2 m depth class. Small lakes and the outer edges of larger lakes were placed to the 0-1 m depth class. The area of lakes in each of these depth classes was tallied and added to the totals derived from the historical Delta DEM.

7.2 Depth by length

Methods used to calculate the linear extent of channels based on their thalweg depths differed for the historical and modern analyses. Historical thalweg depths were generated by segmenting the historical thalweg polyline (developed with/for the historical Delta DEM) at intervals of approximately 100 m and then intersecting the segments with the historical Delta DEM. Each 100 m segment was attributed with the average bed-elevation associated with the raster cells it crossed. These thalweg bed-elevations were converted to water depths using the methods/table described in Section 7.

Modern thalweg depths were generated by intersecting the trimmed modern channel width transects (see Section 6) with the clipped modern bathymetry raster (see Section 4.2) and attributing each transect with the minimum encountered cell value (i.e., the lowest bed-elevation). This is akin to taking the minimum value from a channel cross-section. Minimum bed-elevations were then attributed to the modern channel polyline segments and converted to water-depths using the methods/table described in Section 7.

8. CHANNEL ADJACENCY

Channel adjacency was determined from the habitat type layers by extracting habitat types associated with open water or aquatic habitat (for historical: 'Fluvial low order channel,' 'Fluvial mainstem channel,' 'Tidal low order channel,' 'Tidal mainstem channel,' 'Non-tidal intermittent pond/lake,' 'Non-tidal perennial pond/lake,' and 'Tidal intermittent pond/lake,'; for modern: 'Water') and intersecting the resulting layer with all other habitat types. The output of this operation is a polyline that traces the locations where open water touches other habitat types (the "shoreline"), and includes all of the attributes of the adjacent habitat type polygons.

Also included as open water when generating the historical shoreline layer were the historical channel polylines (which, due to their size, were not represented a polygons in the habitat types layer). A buffer of 5 m was applied to each side of the polylines to give the features an area. Before shorelines were generated, the new open water polygons were incorporated into the habitat layer with ArcGIS's 'Erase' and 'Merge' tools. Shorelines were not generated for possibly exhumed channels (as marked in the channel polylines "Notes" field).

The shoreline layer was used to determine marsh-open water edge length (page 44-45). For this analysis, we selected reaches where the shoreline habitat type was either 'Tidal freshwater emergent wetland' or 'Non-tidal freshwater emergent wetland' (for the historical analysis) or 'Freshwater emergent wetland' or 'Willow-marsh complex' (for the modern analysis). These selections were symbolized by the size of the contiguous marsh polygon they were associated with. Contiguous marsh polygons (which differ from marsh "patches"; see Section 10) were generated by dissolving polygons with the marsh habitat types listed above using the 'Dissolve' tool in ArcGIS 10.1. The sizes of these polygons were attributed to the shorelines with a spatial join.

To assign shoreline data to the channel polylines, channel polylines were segmented at 100 m intervals and given the attributes (via a spatial join) of the nearest shoreline feature. Channels bordered on each side by different habitat types only received attributes from the nearest shoreline feature. We used these methods to determine which dendritic channels were adjacent to marsh (see page 46-47). For this analysis, we considered marsh to be polygons with the habitat types 'Tidal freshwater emergent wetland' or 'Non-tidal freshwater emergent wetland' (for the historical analysis) or 'Freshwater emergent wetland' or 'Willow-marsh complex' (for the modern analysis).

9. LOOPED AND DENDRITIC CHANNELS

We classified tidal channel reaches as either "looped" or "dendritic." Looped channels are interconnected, generally large distributary reaches that delineate the Delta islands and can be thought of as forming circular networks connecting back to the tidal source. They are sometimes referred to as "mainstem and subsidiary channels"[38] or "through-flow channels."[39] Dendritic channels, alternatively, are terminal sloughs that eventually dead-end and do not connect on both ends to the larger network. The term "dendritic" is derived from the typical form of historical terminal sloughs—branching, tree-like networks that terminated in wetlands and resembled dendrites. These sloughs generally drained (and were formed by) tidally introduced water, rather than runoff from associated wetlands and uplands.[40] Although terminal, dead-end sloughs do not always have the branched form today, we still refer to them as "dendritic." These channels have also been referred to as "branching dead-end channel networks,"[41] "backwater tidal sloughs,"[42] "tidal creeks,"[43] and "blind channels."[44]

Ultimately, a channel reach was considered "looped" if it was (1) tidal and (2) connected to the tidal source (the Delta mouth) via two independent and non-overlapping paths (Figure 4). Tidal channel reaches accessible from the tidal source by only one non-overlapping path were considered "dendritic." Classification was carried out manually within ArcGIS. For the historical channel polyline network, tidal channels were selected using the layer's "tidal_status" field, which classified channels as either "tidal" or "fluvial." Since most channels within the study area were at least somewhat influenced by both tidal and fluvial processes, Whipple et al. (2012) classified historical channel reaches by their probable hydrology (instead of by the dominant physical process). Specifically, a channel reach was classified as "tidal" if it likely experienced bidirectional (tidal) flow during spring tides in times of low river stages (even though the primary processes that formed and maintained the channel could be fluvial). "Fluvial" reaches—those upstream of the limit of

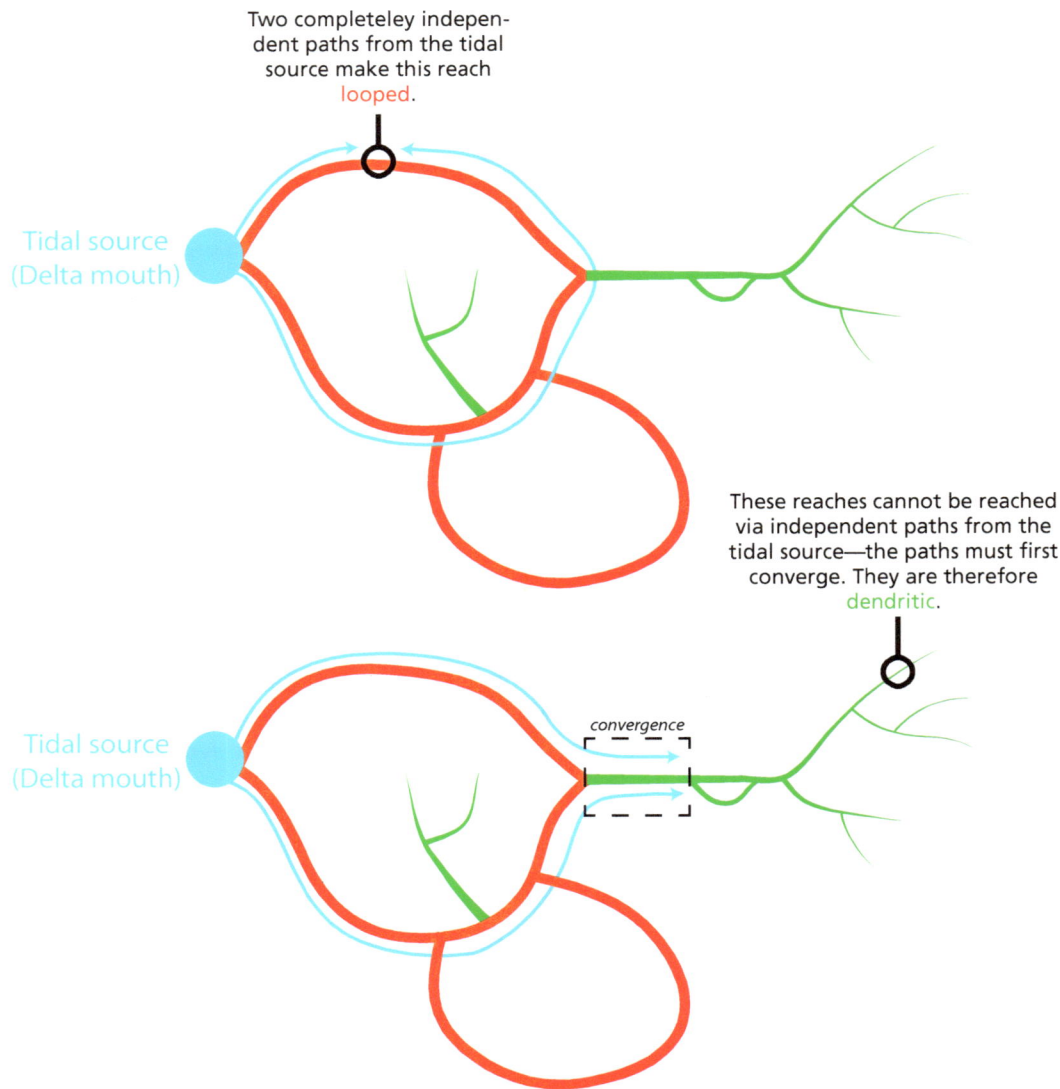

Two completeley independent paths from the tidal source make this reach looped.

Tidal source (Delta mouth)

These reaches cannot be reached via independent paths from the tidal source—the paths must first converge. They are therefore dendritic.

Tidal source (Delta mouth)

convergence

Figure 4. Classifying dendritic and looped channels.

bidirectional flow—were not classified as either dendritic or looped. To identify tidal reaches in the modern network using this definition, we drew from the work of Cavallo et al. (2012) who identified the locations along the Sacramento, San Joaquin, and Mokelumne river systems where bidirectional flows rapidly give way to unidirectional flows under multiple flow regimes.[45] Channel reaches upstream of these transition points under the authors' low-flow scenarios were considered fluvial and not classified as either dendritic or looped.

For both the historical and modern analyses, channel reaches were excluded from the looped/dendritic channel analysis if they lacked a direct, perennial connection (through the larger network) to the Delta mouth (and therefore to the tidal source). This was determined in ArcGIS with a recursive spatial selection that identified intersecting

reaches extending outwards/upstream from the downstream-most channel reach at the Delta mouth. This rule excluded most upland intermittent streams, many of the possibly exhumed and disconnected channels mapped in the historical south Delta,[46] and channels in the modern Delta separated from the tides by levees, weirs, and other barriers (often identified with supplemental information).

Tidal reaches that ultimately connect upstream to perennial fluvial systems were not classified as either dendritic or looped between the perennial fluvial reaches upstream and where they first become looped channels downstream. This rule prevented major rivers like the Sacramento, San Joaquin, and Mokelumne (which are mostly tidal within the study extent) from being lumped with the true dendritic channels that terminate within the study extent.

One final exception to the classification rules described above was made for the large channels bordering Liberty Island in the modern network. Although only one independent path from the tidal source exists for these reaches (paths into the area must converge at the single access point west of the base of the Sacramento Deepwater Ship Channel), they were deemed functionally looped due to their form (a circular path around the former extent of Liberty Island via the "Stair Step" channel) and high local wind wave energy.

10. INUNDATION

10.1 Historical inundation

For the historical Delta, areas regularly subject to inundation were derived from the map of historical habitat types, which were defined, at least in part, by their typical hydrology (Whipple et al. 2012). Areas mapped as 'Tidal freshwater emergent wetland' were classified for the inundation analysis as areas of "tidal inundation"; 'Non-tidal freshwater emergent wetlands' and 'Willow thickets' were classified as areas of "seasonal long-duration flooding"; and 'Vernal pool complex,' 'Wet meadow/seasonal wetland,' and 'Alkali seasonal wetland complex' were classified as areas of "seasonal short-term flooding." Areas mapped as 'Tidal mainstem channel,' 'Fluvial mainstem channel,' 'Tidal low order channel,' 'Fluvial low order channel,' 'Freshwater pond or lake,' and 'Freshwater intermittent pond or lake' were classified as "ponds, lakes, channels, & flooded islands."

The methods described above were further developed in the following ways.

(1) The area mapped by Whipple et al. (2012) as 'Tidal freshwater emergent wetland' (and thus classified as an area of "tidal inundation") represented the area "wetted or inundated by spring tides at low river stages."[47] To distinguish the smaller portion of this area that experienced daily tidal inundation, we relied on the available historical data and best professional judgment.[48] Ultimately, the mapped extent of daily tidal inundation (~76,500 ha) corresponds well with estimates of this area identified in historical records. Most early accounts state that approximately 200,000 acres (80,940 ha) or less were regularly overflowed by "ordinary" tides (i.e., daily high tides).[49] A more specific calculation from an early engineering report states that roughly 160,000 acres (64,750 ha) were "subject to inundation at each high tide, twice in twenty-four hours."[50]

(2) The tidal portion of the lower Yolo Basin, which was only inundated during spring tides (north of the area determined to be inundated daily) was classified both as an area of "tidal inundation" and as area of "seasonal long-duration flooding." This area is displayed on the maps as "seasonal long-duration flooding" during winter and spring and as "tidal inundation" during fall and summer. In the charts on pages 40-41 and 61, the area is included in both categories.

Information on the depth, timing, and duration of each inundation type was derived from Whipple et al. (2012) and other supplemental sources.[51]

10.2 Modern inundation

Since the modern habitat type dataset does not distinguish between tidal and non-tidal freshwater emergent wetland, a proxy was used to define modern areas of tidal inundation. Specifically, areas were assigned the "tidal inundation" classification if they were mapped as either 'Freshwater emergent wetland' or 'Willow-marsh complex,' were adjacent to open water, and fell within the historical extent of tidal marsh. Additional areas of modern inundation were identified, mapped, and classified after conducting a literature search and consulting with regional experts. The extent of the "seasonal short-term flooding" in the Yolo Bypass, for example, was digitized by Sommer et al. (2004) from aerial photographs of the flooding that took place in January 1998. The extent of the Cosumnes River floodplain (also classified as "seasonal short-term flooding") was digitized by SFEI staff from a map of the upper and lower floodplain.[52] We recognize that other areas of the modern Delta may experience inundation, but we only digitized areas identified by the LIT.

11. IDENTIFYING MARSH PATCHES

Historical marsh patches were created from historical habitat type polygons classified as either 'Tidal freshwater emergent wetland' or 'Nontidal emergent wetland'; modern marsh patches were created from modern habitat type polygons classified as either 'Freshwater emergent wetland' or 'Willow-marsh complex.' In the GIS, discrete marsh polygons were aggregated and considered part of a single patch if they were located within 60 m of one another. Groups of polygons separated by less than this distance were identified and aggregated using ArcGIS's 'Aggregate Polygons' tool and assigned unique patch IDs. Multipart feature layers delineating marsh patches (for both the historical and modern Delta) were generated for further analysis (the "patch layers").

The 60 m threshold for grouping marsh polygons was taken from a rule set for defining resident intertidal rail patches developed by Collins and Grossinger (2004), which was based on the best available data on rail habitat affinities and dispersal distances.[11] In the absence of more specific data, we made the assumption that the rules developed for defining intertidal rail patches in the South Bay (primarily for California Clapper Rails, which are not generally found in the Delta) are broadly applicable to the Delta's freshwater (and often non-tidal) marshes/species. Unlike Grossinger and Collins (2004), however, our analysis only considered roads and levees as dispersal barriers if the width of these features (as mapped in the habitat type layers) exceeded the 60 m threshold. It is worth noting that this model of a binary landscape (marsh and non-marsh) simplifies the complexities of how species interact with their surroundings. It necessarily assumes that all patches of marsh are equally suitable for rails, that the routes of travel between patches are linear, and that the only barrier to rail movement is distance.[53]

12. MARSH PATCH SIZE

The size of individual marsh patches was determined with ArcGIS. In addition to determining the size of each patch, we also identified the number and distribution of "large" marsh patches, where "large" was defined based on functionality for marsh bird support. For the purposes of this analysis, a marsh patch was considered "large" if it had an area greater than or equal to 100 ha. This threshold is based on (1) regression models indicating a significant negative correlation between California Black Rail presence and distance to the nearest marsh greater than or equal to 100 ha[54] and (2) research that found that California Clapper Rail densities decrease in patches <100 ha.[55]

13. MARSH CORE VS. EDGE

For the purpose of this analysis, core area index is defined as the percent of a marsh patch's total area that is greater than 50 m from the patch's edge. The core area of each marsh patch was identified in ArcGIS using the 'Buffer' tool with an internal linear buffer distance of 50 m. This distance is based on research indicating a significant positive relationship between California Black Rail presence and marsh core area (defined as >50 m from marsh edge).[56]

14. MARSH NEAREST LARGE NEIGHBOR DISTANCE

Nearest large neighbor distance (NLND) was determined with ArcGIS's 'Generate Near Table' tool, which calculated the linear distance of each marsh patch to the nearest "large" neighboring marsh patch (>100 ha, see Section 12). Large patches themselves were assigned a NLND of 0 m. This metric is supported by research indicating a significant negative relationship between California Black Rail presence and distance to nearest 100 ha marsh.[57]

15. IDENTIFYING RIPARIAN HABITAT PATCHES

Historical riparian patches (here meaning woody riparian habitat patches) were created from historical habitat type polygons clas-

sified as either 'Valley foothill riparian' or 'Willow riparian scrub or shrub.' Modern riparian patches were created from modern habitat type polygons classified as either 'Valley foothill riparian' or 'Willow riparian scrub or shrub,' but also from some polygons ultimately classified as 'Managed wetland' (where the original classification was either 'Valley foothill riparian' or 'Willow riparian scrub or shrub'—see Section 2.4.2). Since, for this analysis, vegetation type and structure were deemed to be more important characteristics than hydrology, riparian habitat type polygons were included whether or not they were deemed hydrologically connected (see Section 2.4.3 for further explanation—this stands in contrast to the riparian width analyses, which exclude hydrologically disconnected riparian habitat polygons).

In the GIS, discrete woody riparian polygons were aggregated and considered part of a single patch if they were located within 100 m of one another. The 100 m threshold for grouping riparian polygons is based on the typical maximum gap crossing distance of dispersing songbirds, as determined by the best professional judgment of regional experts.[58] Groups of polygons separated by less than this distance were identified and aggregated using ArcGIS's 'Aggregate Polygons' tool and assigned unique patch IDs. Multipart feature layers delineating woody riparian habitat patches (for both the historical and modern Delta) were generated for further analysis (the "patch layers"). The size of individual woody riparian habitat patches (and total patch size distribution) for both the historical and modern Delta was determined using these layers with simple ArcGIS table summaries.

As was the case when defining marsh habitat patches, it is worth noting that this model of a binary landscape (woody riparian habitat and non-woody riparian habitat) simplifies the complexities of how species interact with their surroundings. It makes the assumption that all patches of woody riparian habitat are equally suitable for riparian wildlife, that the routes of travel between patches are linear, and that the only barrier to movement is distance.[59]

The thresholds defining woody riparian patch size bins used to assess patch size distribution use a geometric progression starting at 20 ha and multiplying by a common ratio of four. These bins result in thresholds at 20 ha and 80 ha, both of which have apparent ecological significance for Western Yellow-billed Cuckoos. For nesting cuckoos in California, researchers characterize willow-cottonwood patches >80 ha in size as "optimal" and set 20 ha as the minimum threshold for "marginal" habitat suitability.[60] Below this area (<20 ha for mesquite habitat and <15 ha for willow-cottonwood habitat), patches become "unsuitable." The size thresholds of the larger bins do not have specific ecological justifications.

15.1 Estimating the area of unmapped willow-fern swamps

Scattered "clumps" or "patches" of willows are known to have occurred within the tule marshes of many central Delta islands, adding a dimension of woody vertical structure to the freshwater emergent wetland plain.[61] Although not strictly a "riparian" habitat type (the willow patches were not limited to channel banks and are thought to have occurred across central Delta islands where tidal processes were dominant), we quantified the area of this vegetation community since it offers the taller woody vegetation structure that is an important component of many of the functions provided by riparian habitat (sometimes independent of actual hydrological connection to fluvial systems). While "willow-fern swamp" vegetation community was described in detail by Whipple et al. (2012), it was not mapped as a unique habitat type and was instead considered part of the 'Tidal freshwater emergent wetland' habitat type. We estimated the historical area of willow-fern swamp using a historical map made in 1850 by Charles Gibbes[62] and some of the general conclusions drawn from other historical sources by Whipple et al. (2012). In the *Sacramento-San Joaquin Delta Historical Ecology Investigation*, Whipple et al. (2012) determined that willow-fern swamps were most common within Sherman, Bradford, Webb, Venice, and Mandeville islands, and indicated the extent of 'Tidal freshwater emergent wetland' over which the vegetation community is thought to have occurred (Figure 5).[63] We estimated the number individual willow fern-swamp patches in the central Delta by multiplying this area (31,570 ha) by the willow patch density mapped by Gibbes in 1850 (0.007 patches/ha; sampled from a georeferenced version of the map in ArcGIS). The estimated area of willow-fern swamp habitat was determined by multiplying the estimated number of patches (221) by the average patch size (16 ha, determined by measuring the area of a random sample of 35 patches drawn by Gibbes; SD = 12 ha). With this simple operation (and all of its inherent assumptions), we estimated that there were approximately 3,500 ha of willow-fern swamp habitat in the historical central Delta.

16. RIPARIAN HABITAT WIDTH

For both the historical and modern Delta, we sought to visualize and quantify the length of riparian habitat (defined here as woody riparian habitat) based on the riparian habitat's width. We measured historical and modern riparian habitat widths by casting transects

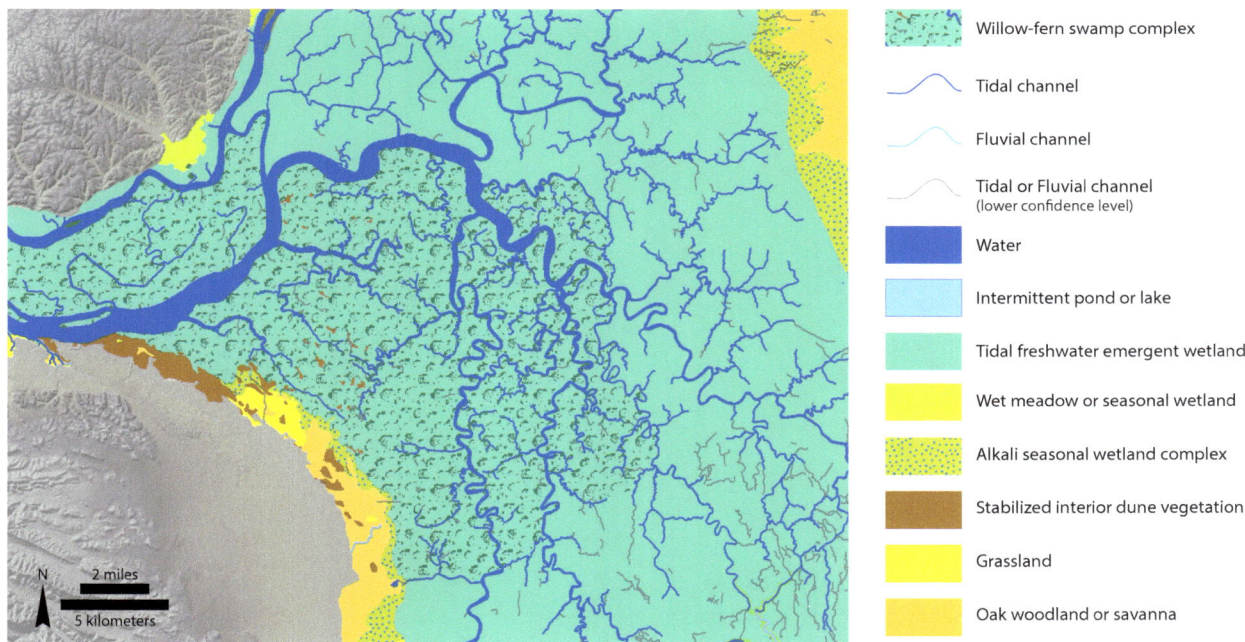

Figure 5. The generalized extent of willow-fern swamp complex (shown by the dark, clumped tree symbol) as mapped by Whipple et al. (2012). The extent was determined from various sources, none of which explicitly described the boundaries of this wetland community. Actual boundaries were likely indiscernible, as the presence of willows within the islands gradually became less prevalent moving away from this mapped area. In this report, we use the approximate extent and data on the density and size of patches sampled from an 1850 map by Charles Gibbes to estimate the total area of willow-fern swamp habitat.

perpendicular to modified channel centerlines and then trimming the transects at the edges of riparian habitat polygons (a method similar to/adapted from our analysis of channel width; see Section 6). The nature of the historical and modern datasets required two different (although generally similar) methods to determine riparian habitat width. These methods are described in detail below.

16.1 Historical riparian habitat width

For the historical layers, riparian areas classified as either 'Valley foothill riparian' or 'Willow riparian scrub or shrub' were extracted from the historical habitat types dataset and merged with adjacent open water polygons. These merged riparian "zones" (including the open water areas) were dissolved, split manually at confluences, and assigned unique identifiers. Next, we generated centerlines for each split riparian habitat zone (from which to cast

perpendicular transects that measure the zone's width at regular intervals). To develop the riparian habitat centerlines, we started with the historical channel polylines, which were modified to adhere to the following rules:

- Riparian centerlines were not drawn for side channels within otherwise contiguous zones of riparian habitat (those that effectively form islands of woody riparian habitat)—these smaller side channels were merged with the larger channel and riparian zone (Figure 6, A and B).

- Riparian centerlines were not drawn for small crevasse splays (Figure 6, C).

- Riparian centerlines were straightened through sinuous areas (Figure 6, D).

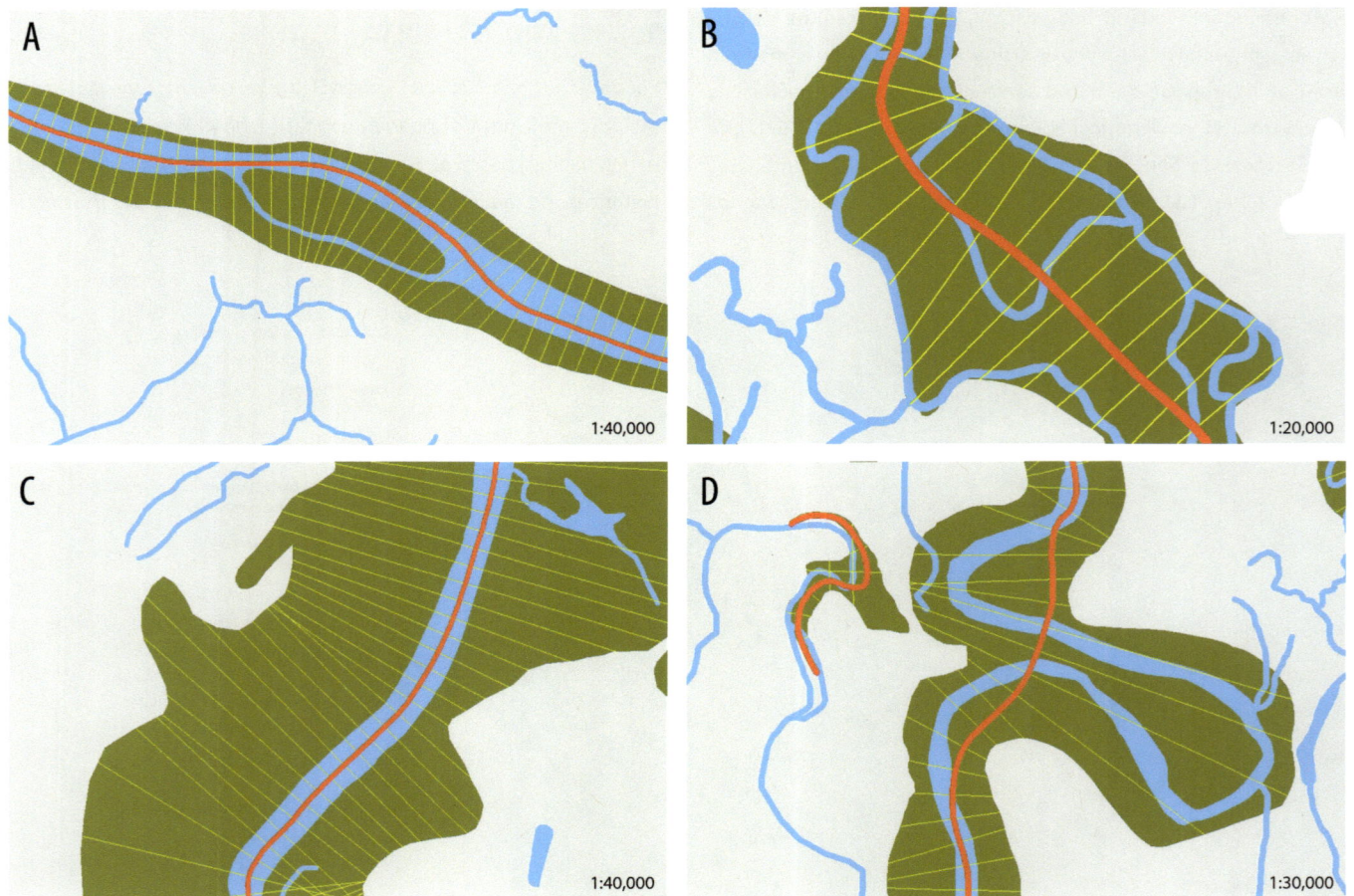

Figure 6. Determining historical riparian width in complicated areas. (A-B) Riparian centerlines were not drawn for side channels within otherwise contiguous zones of riparian habitat. (C) Riparian centerlines were not drawn for small splays (seen here on the left hand side of the woody riparian habitat). (D) Riparian centerlines were straightened through sinuous areas.

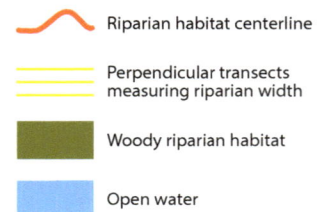

Riparian habitat centerline
Perpendicular transects measuring riparian width
Woody riparian habitat
Open water

- Riparian centerlines were smoothed with a maximum offset of 2 m and generalized by 0.1 to remove sharp angles and to prevent transects from being cast at incorrect angles.

Merged lines were intersected with the riparian habitat zone polygons to associate each line with only one zone. The riparian centerlines were then segmented at 100 m intervals and transects were cast perpendicularly from the centroid of each segment (as determined by the x, y coordinates of its endpoints) 2,000 m in each direction (a distance greater than the maximum width of the woody riparian habitat zone). Transects (containing the unique identifier of the centerline/zone from which they were cast) were then intersected with riparian zone polygons (with the same identifier), thereby trimming the transects to the width of the adjacent woody riparian habitat zone. Since riparian habitat "zones" included both open water and woody riparian habitat, we erased segments of the trimmed transects that intersected open water polygons to determine only the width of the woody riparian habitat. This process was automated with a custom ArcPy script.

16.2 Modern riparian habitat width

Due to the complicated shape and distribution of woody riparian habitat in the modern Delta, we used a second set of methods to determine modern riparian habitat widths. Extensive areas of vegetation in the modern Delta classified as 'Valley foothill riparian' and 'Willow riparian scrub or shrub' are not adjacent to channel features. Since we sought to measure the length/width of riparian habitat along linear zones of open water, we only counted the width of modern woody riparian habitat if it was deemed hydrologically connected (see Section 2.4.3 for how we determined the hydrologic connectivity of woody riparian habitat types; see Figure 7 for a map of the modern woody riparian habitat classified as hydrologically "connected" and "disconnected").

To determine modern riparian habitat widths, we cast transects at 100 m intervals from the modern channel polyline dataset (described in Section 3.2) 1,500 m in each direction. To prevent counting woody riparian vegetation located behind artificial levees (and thus disconnected from the linear channel features), transects were intersected with a polyline layer consisting of artificial levee centerlines.[64] Segments of the transects falling on the far side of the levee centerlines (away from the channel) were discarded. Transects were then intersected with the hydrologically connected woody riparian habitat polygons resulting in trimmed transects with lengths equal to the width of the woody riparian habitat polygons. Trimmed transects were edited manually to remove instances of double counting (where transects cast from one channel intersected riparian habitat

associated with another channel). Riparian habitat only contributed to measurements of width if was associated with the channel reach from which the intersecting transect was cast (determined by visual inspection of the riparian habitat polygons, channel centerlines, and transects). Where there were gaps in the riparian habitat, we counted the area on both sides of the gap towards total riparian width (assuming both areas met the above rule), but did not count the width of the gap itself.

Trimmed riparian width transects >100 m and >500 m were selected to display on the historical and modern maps of woody riparian habitat width. The 100 m width threshold is based on the work of Gaines (1974), who found that Western Yellow-billed Cuckoos were only present in riparian habitat patches at least 100 m wide. The 500 m width threshold is based on the work of Kilgo et al. (1998), who found that riparian forest areas at least 500 m wide were necessary to maintain the "complete avian community" in bottomland hardwood forests in South Carolina. These widths largely agree with the findings of Laymon and Halterman (1989), who (based on occupancy and nest predation rates) define riparian habitat <100 m wide as "unsuitable," habitats 100-600 m wide as "marginal" or "suitable," and habitats at least 600 m wide as "optimal" for cuckoo nesting.

17. MARSH-TERRESTRIAL TRANSITION ZONE LENGTH

The marsh-terrestrial transition zone ("t-zone") was identified using the habitat type layers by extracting habitat type polygons considered "marsh" (described below), generating contiguous polygons from these features (without interior borders), and then intersecting these contiguous polygons with all other habitat type polygons. The output of this operation was a polyline that traces the locations where marsh habitats are directly adjacent to other habitat types. We then extracted segments of this polyline associated with terrestrial habitat types (identified below). This new polyline (that traces locations where marsh shares a border with terrestrial habitat types) was deemed the marsh-terrestrial transition zone. The lengths of t-zone polyline segments (for both the historical and modern datasets) were summed by terrestrial habitat type to generate the chart on page 73.

For this analysis, marsh habitat types were 'Tidal freshwater emergent wetland' (historical), 'Non-tidal freshwater emergent wetland' (historical), 'Freshwater emergent wetland' (modern), and 'Willow-marsh complex' (modern). Terrestrial habitat types were 'Valley foothill riparian,' 'Willow riparian scrub or shrub,' 'Willow thicket,' 'Wet meadow and seasonal wetland,' 'Vernal pool complex,' 'Alkali

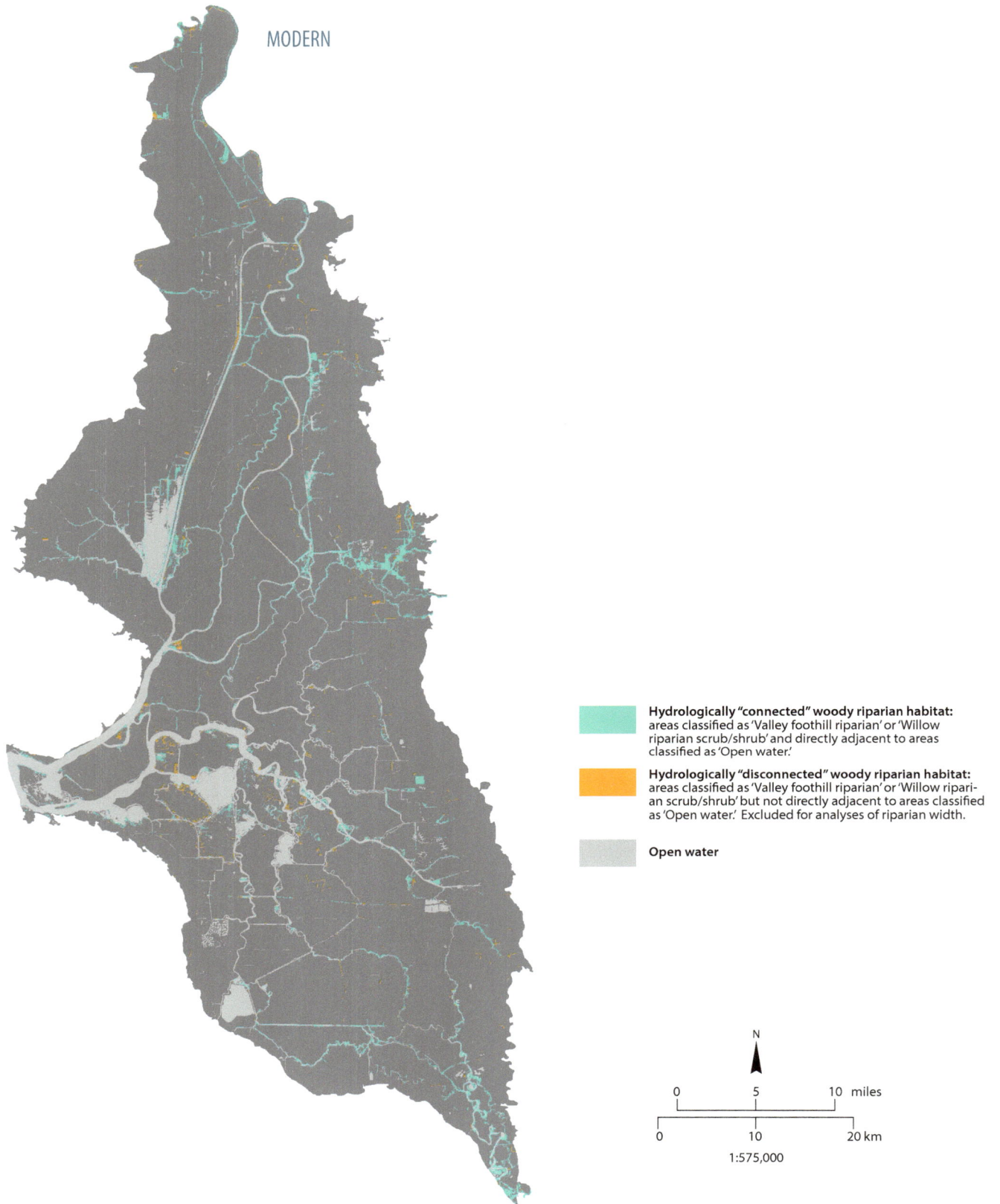

MODERN

(teal)	**Hydrologically "connected" woody riparian habitat:** areas classified as 'Valley foothill riparian' or 'Willow riparian scrub/shrub' and directly adjacent to areas classified as 'Open water.'
(orange)	**Hydrologically "disconnected" woody riparian habitat:** areas classified as 'Valley foothill riparian' or 'Willow riparian scrub/shrub' but not directly adjacent to areas classified as 'Open water.' Excluded for analyses of riparian width.
(gray)	**Open water**

N

0   5   10   miles

0   10   20 km

1:575,000

Figure 7. Hydrologically "connected" and "disconnected" woody riparian habitat in the modern Delta. We distinguished contemporary hydrologically connected woody riparian habitat from hydrologically disconnected woody riparian habitat. Only contiguous woody riparian habitat polygons that shared an edge with open water were deemed hydrologically connected. Although both types were considered when determining woody riparian habitat patch size distribution (pages 64-65), disconnected woody riparian habitat was not considered when calculating riparian width. See Section 16.2 of this chapter for more detailed methods.

seasonal wetland complex,' ' Grassland,' ' Oak woodland and savanna,' and 'Stabilized interior dune vegetation.'

18. CERTAINTY AND LIMITATIONS

18.1 Historical data certainty

Each feature in the historical Delta datasets (habitat type polygons and channels) was assessed for certainty during the mapping process. Whipple et al. (2012) describes this process in the Delta Historical Ecology Investigation:

> Our confidence in a feature's habitat type and presence (interpretation), size, and location was assigned based upon the number of kinds and quality of evidence, accuracy of digitizing source, our experience with the particular aspects of each data source, and by factors such as stability of features on a decadal scale (following standards discussed in Grossinger et al. 2007; [Table 8]). Certainty in tidal status was also included for the channel line layer. In cases where features were likely to have shifted positions over relatively short time periods, we assigned lower certainty for location and size. These attributes provide a way to estimate ranges of uncertainty associated with different locations and kinds of feature or habitat type, and allows subsequent users to assess accuracy [Table 8]. (49)

Using these classifications, the authors were able to assess and roughly quantify the uncertainty associated with the historical mapping:

> Overall, confidence in interpretation and location was fairly high, 64% and 77% respectively. The lower certainty in shape (of each mapped feature) reflects the large areas of habitats, primarily around the perimeter of the Delta, where boundaries were challenging to determine. For the channel lines layer (the network along the polygon channels plus the channels narrower than the polygon minimum mapping width), high interpretation certainty accounted for about 64% of the mapped channel length, with high shape certainty at 59% and high location at 85%. Less than 10% of the area was assigned a low interpretation certainty for either mapping layer. The fourth certainty level standard, tidal interpretation, was only included in the lines layer, where 75% of the channel length was assigned a high certainty level for its tidal interpretation. (89-90)

Mapping certainty varied by habitat type:

> Habitat types with less than 50% of the area assigned with high certainty include alkali seasonal wetland complex, grassland, tidal intermittent pond or lake, vernal pool complex, wet meadow or seasonal wetland, willow riparian scrub or shrub, and willow thicket. Habitat types associated with the highest interpretation certainty tended to be the water bodies and freshwater emergent wetland, given the many sources available confirming these habitat types (e.g., descriptions of tule to identify freshwater emergent wetland). Not surprisingly, the similar summary of the channel line layer shows the larger mainstem channels that are well-established in numerous historical sources with nearly 100% interpretation certainty, while the interpretation of lower order channels was more challenging, mostly due to the difficulties associated with distinguishing the early 1800s channels from the many signatures of ancient channels exposed by exhumed peat in the south Delta. (90-91).

For a full discussion of the uncertainties associated with the historical habitat types and channel datasets, please refer to the *Delta Historical Ecology Investigation*.[65]

Table 8. **Certainty level standards** assigned to each mapped historical feature for the assessment of confidence in interpretation (classification and historical presence), size, location, and tidal status. From Whipple et al. (2012).

Certainty Level	Interpretation	Size	Location	Tidal Status (line features only)
High/ "Definite"	Feature definitely present before Euro-American modification	Mapped feature expected to be 90%-110% of actual feature size	Expected maximum horizontal displacement less than 50 m (150 ft)	Channel bed definitely within or outside tidal range (<3.5 ft elevation)
Medium/ "Probable"	Feature probably present before Euro-American modification	Mapped feature expected to be 50%-200% of actual feature size	Expected maximum horizontal displacement less than 150 m (500 ft)	Channel bed probably within or outside tidal range
Low/ "Possible"	Feature possibly present before Euro-American modification	Mapped feature expected to be 25%-400% of actual feature size	Expected maximum horizontal displacement less than 500 m (1,600 ft)	Channel bed possibly within or outside tidal range (if within, no clear tidal connection)

18.2 Modern data certainty

As a compilation of multiple sources, the modern habitat types layer utilized in this report represents a conglomeration of certainty levels that vary within and between the individual sources. The two primary modern data sources combined in this report each underwent independent assessments of mapping accuracy. For the VegCAMP 2007 Sacramento-San Joaquin River Delta dataset ('CDFG 2007 Delta Vegetation'—the source for 80% of this project's study extent) accuracy was assessed using the fuzzy logic method.[66] The overall accuracy of the map was nearly 89%, while the average accuracy score per vegetation type was 83%. For the Central Valley Riparian Mapping Project Group Level dataset ('CDWR 2012 CVRMP'—the source for 19% of this project's study extent) accuracy was assessed by comparing how photo interpreters (producers) and field surveyors (users) classified the same regions.[67] The overall user's accuracy score averaged 76% and the producer's accuracy averaged 79%.

Some uncertainty was also introduced through the development of the crosswalk used to relate each of the different original classification systems (to each other and to the historical classifications). As noted by Hickson and Keeler-Wolf (2007), "The complexity and uncertainty of such relationships arise not only from independent evolution of classifications, but also from their imprecise definitions, without quantitative rules for proper interpretation. The best crosswalks are those that have been developed with a good understanding of the meaning and definitions of each classification system." By having Todd Keeler-Wolf (an author of the primary modern dataset utilized in this report) assist with the development of this project's crosswalk, we were able to minimize the uncertainty associated with a somewhat subjective process.

Since our modern mapping is from a compilation of sources, it represents a compilation of years. The oldest—the VegCAMP 2007 Sacramento-San Joaquin River Delta dataset—utilized U.S. Geological Survey High Resolution Orthoimagery taken in 2002 and 2005.[68] The most recent source—supplemental polygons digitized by SFEI staff (covering less than 1% of the study extent)—was derived from Bing aerial photos accessed in 2013. The Delta is a continually changing place and there is uncertainty associated with modern classifications that are already outdated at the time of publication; we are aware of at least seven sizeable parcels (including areas mapped as 'Grassland,' 'Wet meadow and seasonal wetland,' and 'Agriculture/Non-native/Ruderal') that have been developed since the modern habitat type datasets were generated.[69]

18.3 Issues of historical and modern data fidelity: comparing apples to apples (or at least to crabapples)

One of the fundamental goals of this report was to ensure that, when making comparisons between the historical and modern landscape, we compared the same things, at the same scale, using the same measurements. Due to the severity of change in the Delta and differences between the historical and modern datasets, this task was far from trivial. In this section we discuss the consequences of differences in historical and modern data resolution. These differences were more or less pronounced depending on the datasets used and the analyses in question. The extent to which we could control for differences in resolution also varied across analyses. While some analyses are affected by differences in data resolution (that increase the uncertainty surrounding specific numbers and the precise magnitude of the measured changes), we do not believe that these differences impact the direction of changes or the overall stories indicated by our analyses.

Generally speaking, the spatial data for the modern Delta has a higher resolution than the spatial data for the historical Delta, but these differences are not always very pronounced and were largely manageable. In our analysis of marsh core area, for example, it was important to make sure that the resolution of non-marsh features within the marsh (which effectively create marsh edge) was similar in the historical and modern datasets. When calculating historical core area ratio, we chose only to include channels mapped as polygons, because their minimum mapping width (15 m; MMW) was comparable to the MMW for water features in the modern dataset (10 m; see Table 1). Although not identical, these MMWs are well within an order of magnitude of one another. While it is true that the slightly lower MMW for water features in the modern dataset increases the amount of modern edge habitat, this difference is insignificant when comparing the core area ratios of the historical and modern marshes: the vast area of largely contiguous historical marsh ensures a higher core area ratio. Similarly, since willow patches in the historical central Delta were not explicitly mapped (due to a lack of data) and instead were lumped into the tidal freshwater emergent wetland classification, we made a concerted effort to do the same for the modern dataset (because the historical lumping effectively decreases marsh edge). This was largely accomplished through the modern data crosswalk (which included areas of marsh and some woody vegetation in the 'Freshwater emergent wetland' category) but also by generating a 'Willow-marsh complex' designation that allowed us to further lump areas of willows with freshwater emergent wetland species into the areas we considered "marsh" when calculating marsh core area ratio (see sections 2.4.1 and 11).

Decisions like this increased fidelity between the historical and modern analyses.

The historical and modern habitat type datasets also utilized different minimum mapping units (MMUs) for areal features—5 ha in the historical dataset and 0.8 ha (for vegetation) in the modern dataset. While these values are within an order of magnitude, their difference is still a concern, because the inclusion of a smaller class of features in the modern dataset that are not included in the historical dataset can increase estimates of patch number and edge length, while decreasing estimates of average patch size. Since outright exclusion of the smallest features in the modern Delta (to match the MMU of the historical Delta) would eliminate a significant proportion of most habitat types and generate unwieldy data gaps, we instead developed methodologies and analyses that minimize/manage the impact of MMU differences and consider here how the differences are likely affecting our results.

One method we used to manage the difference in minimum mapping units was to aggregate individual polygons into patches (see sections 11 and 15; marsh polygons were aggregated if less than 60 m apart, riparian polygons if less than 100 m apart). Small, highly resolved modern features, if proximal to one another, were not counted separately and were effectively lumped to a size above the historical mapping unit. Although small, unmapped areas of marsh certainly existed in the historical Delta, these areas would have had to exist more than 60 m away from a mapped marsh to impact the number of historical patches in our analysis. The same goes for unresolved gaps in the historical marsh—unless these gaps isolated an area of marsh 60 m in all directions, the total number of marsh patches was not affected. Although the process of aggregating polygons into patches minimizes the effects of different patch sizes on our analyses, many modern patches analyzed in Chapter 5 are below the minimum historical mapping unit. To assess the impact of including these patches on our landscape metrics, and the sensitivity of the modern analyses to differences in MMUs, we calculated marsh patch statistics without patches less than 5 ha in size (the historical minimum mapping unit). Doing so yielded a significant decrease in the total number of patches—from 1,211 to 43. Average patch size increased, but perhaps less dramatically—from 4 ha (SD = 24 ha) to 22 ha (SD = 66 ha). It is worth noting that 9 (of 43) historical marsh patches are also below the historical dataset's minimum mapping unit (largely due to study boundary conditions)—enforcing a 5 ha MMU would thus increase historical average patch size from 4,494 ha (SD = 17,956 ha) to 5,682 ha (SD = 20,085 ha). Although removal of the marsh patches less than 5 ha would affect the precise magnitude of change, the direction of change and larger story remain unchanged.

The fidelity of the historical and modern channel polyline datasets is quite high. Both used no minimum mapping width, and channels were digitized wherever evidence of them existed. As described in Section 5, we estimated the length of unmapped historical low-order channels that we expect are comparable in size to the smallest channels visible/digitized in the modern Delta. Although these channels are not explicitly drawn on the map, accounting for their estimated length allowed us to more effectively make comparisons with the modern channel dataset. In both datasets, channels were only digitized around islands larger than 25 ha.

Table 2. Habitat types used to map the historical habitats of the Sacramento-San Joaquin Delta.

Landcover grouping	Habitat type	Description	MSCS NCCP Habitat Types (CALFED 2000c)
Water	Tidal mainstem channel	Rivers, major creeks, or major sloughs forming Delta islands where water is understood to have ebbed and flowed in the channel at times of low river flow. These delineated the islands of the Delta.	Tidal Perennial Aquatic
	Fluvial mainstem channel	Rivers or major creeks with no influence of tides.	Valley Riverine Aquatic
	Tidal low order channel	Dendritic tidal channels (i.e., dead-end channels terminating within wetlands) where tides ebbed and flowed within the channel at times of low river flow.	Tidal Perennial Aquatic
	Fluvial low order channel	Distributaries, overflow channels, side channels, swales. No influence of tides. These occupied non-tidal floodplain environments or upland alluvial fans.	Valley Riverine Aquatic
	Freshwater pond or lake	Permanently flooded depressions, largely devoid of emergent Palustrine vegetation. These occupied the lowest-elevation positions within wetlands.	Tidal Perennial Aquatic, Lacustrine
	Freshwater intermittent pond or lake	Seasonally or temporarily flooded depressions, largely devoid of emergent Palustrine vegetation. These were most frequently found in vernal pool complexes at the Delta margins and also in the non-tidal floodplain environments.	N/A
Freshwater emergent wetland	Tidal freshwater emergent wetland	Perennially wet, high water table, dominated by emergent vegetation. Woody vegetation (e.g., willows) may be a significant component for some areas, particularly the western-central Delta. Wetted or inundated by spring tides at low river stages (approximating high tide levels).	Tidal Freshwater Emergent
	Non-tidal freshwater emergent wetland	Temporarily to permanently flooded, permanently saturated, freshwater non-tidal wetlands dominated by emergent vegetation. In the Delta, occupying upstream floodplain positions above tidal influence.	Non-tidal Freshwater Permanent Emergent
Willow thicket and riparian forest	Willow thicket	Perennially wet, dominated by woody vegetation (e.g., willows), emergent vegetation may be a significant component, generally located at the "sinks" of major creeks or rivers as they exit alluvial fans into the valley floor.	Valley/Foothill Riparian

Wildlife Habitat Relationship (WHR)	Representative types from California Terrestrial Natural Communities (CNDDB 2010)	Cowardin et al. (1979)/ USFWS Riparian Mapping System (USFWS 2009)	Hydrogeomorphic classification (HGM) (Brinson 1993)
Estuarine, Riverine	*Azolla (filiculoides, mexicana)* (Mosquito fern mats) Provisional Alliance (52.106.00), *Stuckenia (pectinata) - Potamogeton* spp. (Pondweed mats) Alliance (52.107.00)	Estuarine subtidal, Estuarine intertidal, Riverine	Riverine wetland, surface flow, unidirectional flow and bidirectional flow
Estuarine, Riverine	*Azolla (filiculoides, mexicana)* (Mosquito fern mats) Provisional Alliance (52.106.00), *Stuckenia (pectinata) - Potamogeton* spp. (Pondweed mats) Alliance (52.107.00)	Estuarine subtidal, Estuarine intertidal, Riverine	Riverine wetland, surface flow, unidirectional flow and bidirectional flow
Estuarine, Riverine	*Azolla (filiculoides, mexicana)* (Mosquito fern mats) Provisional Alliance (52.106.00), *Stuckenia (pectinata) - Potamogeton* spp. (Pondweed mats) Alliance (52.107.00)	Estuarine subtidal, Estuarine intertidal, Riverine	Riverine wetland, surface flow, unidirectional flow and bidirectional flow
Estuarine, Riverine	*Azolla (filiculoides, mexicana)* (Mosquito fern mats) Provisional Alliance (52.106.00), *Stuckenia (pectinata) - Potamogeton* spp. (Pondweed mats) Alliance (52.107.00)	Estuarine subtidal, Estuarine intertidal, Riverine	Riverine wetland, surface flow, unidirectional flow and bidirectional flow
Estuarine, Lacustrine	*Azolla (filiculoides, mexicana)* (Mosquito fern mats) Provisional Alliance (52.106.00), *Stuckenia (pectinata) - Potamogeton* spp. (Pondweed mats) Alliance (52.107.00), *Nuphar polysepala* (Yellow pond-lily mats) Provisional Alliance (52.110.00)	Lacustrine	Depressional wetland, surface flow and groundwater, vertical fluctuations
N/A	N/A	N/A	Depressional wetland, surface flow and groundwater, vertical fluctuations
Fresh Emergent Wetland	*Schoenoplectus acutus* (Hardstem bulrush marsh) Alliance (52.122.00), *Schoenoplectus californicus* (California bulrush marsh) Alliance (52.114.00), *Typha (domingensis, latifolia)* (Cattail marshes) Alliance (52.050.00), American bulrush marsh (52.111.00), California bulrush marsh (52.114.00), *Juncus effusus* (Soft rush marshes) Alliance (45.561.00), *Juncus articus* (Baltic and Mexican rush marshes) Alliance (45.562.00), *Salix lucida* (Shining willow groves) Alliance (61.204.00), *Eleocharis macrostachya* (Pale spike rush marshes) Alliance (45.230.00)	Estuarine intertidal persistent emergent wetland. Temporarily to seasonally flooded, permanently saturated.	Fringe wetland, surface flow including tidal, bidirectional flow
Fresh Emergent Wetland	*Schoenoplectus acutus* (Hardstem bulrush marsh) Alliance (52.122.00), *Schoenoplectus californicus* (California bulrush marsh) Alliance (52.114.00), *Typha (domingensis, latifolia)* (Cattail marshes) Alliance (52.050.00), *Juncus effusus* (Soft rush marshes) Alliance (45.561.00), *Juncus articus* (Baltic and Mexican rush marshes) Alliance (45.562.00), *Eleocharis macrostachya* (Pale spike rush marshes) Alliance (45.230.00)	Palustrine persistent emergent freshwater wetland. Temporarily to permanently flooded, permanently saturated.	Riverine wetland, surface flow, unidirectional flow
Valley foothill riparian	*Salix gooddingii* Alliance (61.211.00), *Salix laevigata* Alliance (61.205.00), *Salix lasiolepis* Alliance (61.201.00), *Salix lucida* Alliance (61.204.00), *Salix exigua* Alliance (61.209.00), *Cornus sericea* (Red osier thickets) Alliance (80.100.00), *Rosa californica* Alliance (63.907.00), *Acer negundo* (Box-elder forest) Alliance (61.440.00), *Sambucus nigra* (Blue elderberry stands) Alliance	Palustrine forested wetland. Temporarily flooded, permanently saturated. / Riparian scrub/shrub deciduous.	Riverine wetland, surface flow, vertical fluctuations

Table 2 (continued). Habitat types used to map the historical habitats of the Sacramento-San Joaquin Delta.

Landcover grouping	Habitat type	Description	MSCS NCCP Habitat Types (CALFED 2000c)
Willow thicket and riparian forest (continued)	Willow riparian scrub or shrub	Riparian vegetation dominated by woody scrub or shrubs with few to no tall trees. This habitat type generally occupies long, relatively narrow corridors of lower natural levees along rivers and streams.	Valley/Foothill Riparian
	Valley foothill riparian	Mature riparian forest usually associated with a dense understory and mixed canopy, including sycamore, oaks, willows, and other trees. Occupied the supratidal natural levees of larger rivers that were occasionally flooded.	Valley/Foothill Riparian
Seasonal wetland	Wet meadow or seasonal wetland	Temporarily or seasonally flooded, herbaceous communities characterized by poorly-drained, clay-rich soils. These often comprised the upland edge of perennial wetlands.	Natural Seasonal Wetland
	Vernal pool complex	Area of seasonally flooded depressions, characterized by a relatively impermeable subsurface soil layer and distinctive vernal pool flora. These often comprised the upland edge of perennial wetlands.	Natural Seasonal Wetland
	Alkali seasonal wetland complex	Temporarily or seasonally flooded, herbaceous or scrub communities characterized by poorly-drained, clay-rich soils with a high residual salt content. These often comprised the upland edge of perennial wetlands.	Natural Seasonal Wetland
Other upland	Stabilized interior dune vegetation	Vegetation dominated by shrub species with some locations also supporting live oaks on the more stabilized dunes with more well-developed soil profiles.	Inland Dune Scrub
	Grassland	Low herbaceous communities occupying well-drained soils and composed of native forbs and annual and perennial grasses and usually devoid of trees. Few to no vernal pools present.	Grassland
	Oak woodland or savanna	Oak dominated communities with sparse to dense cover (10-65% cover) and an herbaceous understory.	Valley/Foothill Woodland and Forest

Wildlife Habitat Relationship (WHR)	Representative types from California Terrestrial Natural Communities (CNDDB 2010)	Cowardin et al. (1979)/ USFWS Riparian Mapping System (USFWS 2009)	Hydrogeomorphic classification (HGM) (Brinson 1993)
Valley foothill riparian	*Salix gooddingii* Alliance (61.211.00), *Salix laevigata* Alliance (61.205.00), *Salix lasiolepis* Alliance (61.201.00), *Salix lucida* Alliance (61.204.00), *Salix exigua* Alliance (61.209.00), *Cornus sericea* (Red osier thickets) Alliance (80.100.00), *Rosa californica* Alliance (63.907.00), *Acer negundo* (Box-elder forest) Alliance (61.440.00), *Cephalanthus occidentalis* (Button willow thickets) Alliance (63.300.00)	Palustrine forested wetland. Intermittently flooded, seasonally saturated. / Riparian scrub/ shrub deciduous.	Riverine wetland, surface flow, vertical fluctuations
Valley foothill riparian	*Quercus agrifolia* Alliance (71.060.00), *Quercus lobata* Alliance (71.040.00), *Quercus (agrifolia, douglasii, garryana, kelloggii, lobata, wislizeni)* Alliance (71.100.00), *Quercus wislizeni* Alliance (71.080.00), *Juglans hindsii and Hybrids* Special stands (61.810.00), *Salix gooddingii* Alliance (61.211.00), *Salix laevigata* Alliance (61.205.00), *Salix lasiolepis* Alliance (61.201.00), *Salix lucida* Alliance (61.204.00), *Salix exigua* Alliance (61.209.00), *Acer negundo* (Box-elder forest) Alliance (61.440.00), *Cornus sericea* (Red osier thickets) Alliance (80.100.00), *Rosa californica* Alliance (63.907.00), *Platanus racemosa* Alliance (61.310.00), *Populus fremontii* Alliance (61.130.00), *Cephalanthus occidentalis* (Button willow thickets) Alliance (63.300.00)	Palustrine forested wetland. Intermittently flooded, seasonally saturated. / Riparian forested deciduous	Riverine wetland, surface flow, vertical fluctuations
Wet meadow	*Lasthenia californica - Plantago erecta - Vulpia microstachys* (California goldfields-dwarf plantain-six-weeks fescue flower fields) Alliance (44.108.00), *Elymus triticoides* (Creeping rye grass turfs) Alliance (41.080.00), *Ambrosia psilostachya* (Western ragweed meadows) Alliance (33.065.00), *Lotus purshianus* (Spanish clover fields) Provisional Herbaceous Alliance (52.230.00), *Juncus effusus* (Soft rush marshes) Alliance (45.561.00), *Juncus articus* (Baltic and Mexican rush marshes) Alliance (45.562.00)	Palustrine emergent wetland. Temporarily to seasonally flooded, seasonally saturated.	Depressional wetland, surface flow and groundwater, vertical fluctuations
Annual grassland	*Lasthenia fremontii - Downingia (bicornuta)* (Fremont's goldfields - Downingia vernal pools) Alliance (42.007.00), *Eryngium aristulatum* Alliance (42.004.00)	Palustrine nonpersistent emergent wetland.	Depressional wetland, surface flow and precipitation, vertical fluctuations
Alkali desert scrub	*Cressa truxillensis - Distichlis spicata* (Alkali weed - Salt grass playas and sinks) Alliance (46.100.00), *Lasthenia fremontii - Distichlis spicata* (Fremont's goldfields - Saltgrass alkaline vernal pools) Alliance (44.119.00), *Allenrolfea occidentalis* (Iodine bush scrub) Alliance (36.120.00), *Sporobolus airoides* (Alkali sacaton grassland) Alliance (41.010.00), *Elymus triticoides* (Creeping rye grass turfs) Alliance (41.080.00), *Frankenia salina* (Alkali heath marsh) Alliance (52.500.00)	Palustrine emergent saline wetland. Temporarily to seasonally flooded, seasonally to permanently saturated.	Depressional wetland, surface flow and precipitation, vertical fluctuations
Coastal scrub	*Lupinus albifrons* (Silver bush lupine scrub) Alliance (32.081.00), *Baccharis pilularis* (Coyote brush scrub) Alliance (32.060.00), *Lotus scoparius* (Deer weed scrub) Alliance (52.240.00)	N/A	N/A
Annual grassland, Perennial grassland	*Lasthenia californica - Plantago erecta - Vulpia microstachys* (California goldfields - Dwarf plantain - Six-weeks fescue flower fields) Alliance (44.108.00), *Elymus triticoides* (Creeping rye grass turfs) Alliance (41.080.00), *Nassella pulchra* Alliance (41.150.00), *Eschscholzia (californica)* (California poppy fields) Alliance (43.200.00), *Amsinckia* (Fiddleneck fields) Alliance (42.110.00), *Plagiobothrys nothofulvus* (Popcorn flower fields) Alliance (43.300.00)	N/A	N/A
Valley oak woodland, Blue oak woodland, Coastal oak woodland	*Quercus agrifolia* Alliance (71.060.00), *Quercus lobata* Alliance (71.040.00), *Quercus (agrifolia, douglasii, garryana, kelloggii, lobata, wislizeni)* Alliance (71.100.00), *Quercus wislizeni* Alliance (71.080.00), *Quercus douglasii* Alliance (71.020.00)	N/A	N/A

Table 3. Crosswalk for the datasets used to generate a complete modern Delta habitat type map.

Crosswalked habitat type	Original classifications, by dataset (with relevant field)		
	CDFG 2007 Delta Vegetation ("MAPUNIT")	CDWR 2012 CVRMP ("DELTAVEG" [priority] or "NVCSNAME")	WWR 2013 CSCCA Natural Communities & CDWR 2013 BDCP Natural Communities ("SAIC_TYPE")
Agriculture/ Non-native/ Ruderal	*Acacia - Robinia*	Agriculture	Agricultural
	Agriculture	Californian warm temperate marsh/seep	
	Eucalyptus	Exotic Vegetation Stands	
	Exotic Vegetation Stands	Giant Cane (*Arundo donax*)	
	Giant Cane (*Arundo donax*)	Intermittently or Temporarily Flooded Deciduous Shrublands	
	Horsetail (*Equisetum* spp.)	Introduced North American Mediterranean woodland and forest	
	Intermittently or Temporarily Flooded Deciduous Shrublands	Mediterranean California naturalized annual and perennial grassland	
	Lepidium latifolium - Salicornia virginica - Distichlis spicata	Pampas Grass (*Cortaderia selloana - C. jubata*)	
	Microphyllous Shrubland	Ruderal Herbaceous Grasses & Forbs	
	Pampas Grass (*Cortaderia selloana - C. jubata*)	Sparsely or Unvegetated Areas; Abandoned orchards	
	Perennial Pepperweed (*Lepidium latifolium*)		
	Poison Hemlock (*Conium maculatum*)		
	Ruderal Herbaceous Grasses & Forbs		
	Sparsely or Unvegetated Areas; Abandoned orchards		
	Tobacco brush (*Nicotiana glauca*) mapping unit		
Alkali seasonal wetland complex	Alkali Heath (*Frankenia salina*)	Pickleweed (*Salicornia virginica*)	
	Alkaline vegetation mapping unit	Saltgrass (*Distichlis spicata*)	
	Allenrolfea occidentalis mapping unit	Southwestern North American salt basin and high marsh	
	Distichlis spicata - Salicornia virginica		
	Frankenia salina - Distichlis spicata		
	Juncus bufonius (salt grasses)		
	Pickleweed (*Salicornia virginica*)		
	Salicornia virginica - Cotula coronopifolia		
	Salicornia virginica - Distichlis spicata		
	Salt scalds and associated sparse vegetation		
	Saltgrass (*Distichlis spicata*)		
	Suaeda moquinii - (Lasthenia californica) mapping unit		
Freshwater emergent wetland	American Bulrush (*Scirpus americanus*)	Arid West freshwater emergent marsh	Tidal Brackish Emergent Wetland
	Broad-leaf Cattail (*Typha latifolia*)	Mixed *Scirpus* / Submerged Aquatics (*Egeria-Cabomba-Myriophyllum* spp.) complex	Tidal Freshwater Emergent Wetland
	California Bulrush (*Scirpus californicus*)	*Scirpus acutus - Typha angustifolia*	
	Common Reed (*Phragmites australis*)	*Scirpus acutus* Pure	
	Hard-stem Bulrush (*Scirpus acutus*)		

Crosswalked habitat type	Original classifications, by dataset (with relevant field)		
	CDFG 2007 Delta Vegetation ("MAPUNIT")	CDWR 2012 CVRMP ("DELTAVEG" [priority] or "NVCSNAME")	WWR 2013 CSCCA Natural Communities & CDWR 2013 BDCP Natural Communities ("SAIC_TYPE")
Freshwater emergent wetland	Mixed *Scirpus* / Floating Aquatics (*Hydrocotyle - Eichhornia*) Complex Mixed *Scirpus* / Submerged Aquatics (*Egeria-Cabomba-Myriophyllum* spp.) complex Mixed *Scirpus* Mapping Unit Narrow-leaf Cattail (*Typha angustifolia*) *Polygonum amphibium* *Scirpus acutus - (Typha latifolia) - Phragmites australis* *Scirpus acutus - Typha angustifolia* *Scirpus acutus* Pure *Scirpus acutus -Typha latifolia* *Scirpus californicus - Eichhornia crassipes* *Scirpus californicus - Scirpus acutus* *Scirpus* spp. in managed wetlands Smartweed *Polygonum* spp. - Mixed Forbs *Typha angustifolia - Distichlis spicata*		
Grassland	*Bromus diandrus - Bromus hordeaceus* California Annual Grasslands - Herbaceous Creeping Wild Rye Grass (*Leymus triticoides*) Italian Rye-grass (*Lolium multiflorum*) *Lolium multiflorum - Convolvulus arvensis* Tall & Medium Upland Grasses	California annual forb/grass vegetation California Annual Grasslands - Herbaceous Italian Rye-grass (*Lolium multiflorum*)	Grassland
Interior dune scrub	*Lotus scoparius* - Antioch Dunes *Lupinus albifrons* - Antioch Dunes		
Managed wetland			Managed Wetland
Urban/Barren	Levee Rock Riprap Urban Developed - Built Up	Barren Urban Urban Developed - Built Up	Developed
Valley foothill riparian	Black Willow (*Salix gooddingii*) - Valley Oak (*Quercus lobata*) restoration Coast Live Oak (*Quercus agrifolia*) Fremont Cottonwood (*Populus fremontii*) Hinds walnut (*Juglans hindsii*) Oregon Ash (*Fraxinus latifolia*) *Quercus lobata - Acer negundo* *Quercus lobata - Alnus rhombifolia (Salix lasiolepis - Populus fremontii - Quercus agrifolia)*	Black Willow (*Salix gooddingii*) Californian broadleaf forest and woodland Central and south coastal California seral scrub Coast Live Oak (*Quercus agrifolia*) Fremont Cottonwood (*Populus fremontii*) *Quercus lobata - Alnus rhombifolia (Salix lasiolepis - Populus fremontii - Quercus agrifolia)* *Quercus lobata - Fraxinus latifolia*	Valley/Foothill Riparian

	Original classifications, by dataset (with relevant field)		
Crosswalked habitat type	CDFG 2007 Delta Vegetation ("MAPUNIT")	CDWR 2012 CVRMP ("DELTAVEG" [priority] or "NVCSNAME")	WWR 2013 CSCCA Natural Communities & CDWR 2013 BDCP Natural Communities ("SAIC_TYPE")
Valley foothill riparian	*Quercus lobata - Fraxinus latifolia* *Quercus lobata / Rosa californica* (*Rubus discolor - Salix lasiolepis / Carex* spp.) Restoration Sites *Salix gooddingii - Populus fremontii - (Quercus lobata-Salix exigua-Rubus discolor)* *Salix gooddingii - Quercus lobata* / Wetland Herbs Temporarily or Seasonally Flooded - Deciduous Forests Tree-of-Heaven (*Ailanthus altissima*) Valley Oak (*Quercus lobata*) Valley Oak (*Quercus lobata*) restoration	*Quercus lobata / Rosa californica* (*Rubus discolor - Salix lasiolepis / Carex* spp.) *Salix gooddingii - Populus fremontii - (Quercus lobata-Salix exigua-Rubus discolor)* *Salix gooddingii - Quercus lobata* / Wetland Herbs Southwestern North American riparian evergreen and deciduous woodland Valley Oak (*Quercus lobata*)	
Vernal pool complex	Vernal Pools	Californian mixed annual/perennial freshwater vernal pool/swale/plain bottomland	Vernal Pool Complex
Water	Algae Brazilian Waterweed (*Egeria - Myriophyllum*) Submerged Floating Primrose (*Ludwigia peploides*) Generic Floating Aquatics *Hydrocotyle ranunculoides* *Ludwigia peploides* Milfoil - Waterweed (generic submerged aquatics) Pondweed (*Potamogeton* sp.) Shallow flooding with minimal vegetation at time of photography Tidal mudflats Water Water Hyacinth (*Eichhornia crassipes*)	Algae Brazilian Waterweed (*Egeria - Myriophyllum*) Submerged Generic Floating Aquatics Riverine Water Western North American Freshwater Aquatic Vegetation	Alkali Seasonal Wetland Complex Non-Tidal Perennial Aquatic Tidal Perennial Aquatic
Wet meadow/ Seasonal wetland	*Distichlis spicata* - Annual Grasses *Distichlis spicata* - Juncus balticus Intermittently Flooded Perennial Forbs Intermittently or temporarily flooded undifferentiated annual grasses and forbs *Juncus balticus* - meadow vegetation Managed alkali wetland (Crypsis)	*Distichlis spicata* - Annual Grasses Intermittently or temporarily flooded undifferentiated annual grasses and forbs Naturalized warm-temperate riparian and wetland Rabbitsfoot grass (*Polypogon maritimus*) Seasonally Flooded Grasslands Seasonally flooded undifferentiated annual grasses and forbs	Other Natural Seasonal Wetland

Crosswalked habitat type	Original classifications, by dataset (with relevant field)		
	CDFG 2007 Delta Vegetation ("MAPUNIT")	CDWR 2012 CVRMP ("DELTAVEG" [priority] or "NVCSNAME")	WWR 2013 CSCCA Natural Communities & CDWR 2013 BDCP Natural Communities ("SAIC_TYPE")
Wet meadow/ Seasonal wetland	Managed Annual Wetland Vegetation (Non-specific grasses & forbs) Rabbitsfoot grass (*Polypogon maritimus*) Seasonally Flooded Grasslands Seasonally flooded undifferentiated annual grasses and forbs Temporarily Flooded Grasslands Temporarily Flooded Perennial Forbs	Temporarily Flooded Perennial Forbs	
Willow riparian scrub/ shrub	*Acer negundo - Salix gooddingii* *Alnus rhombifolia / Cornus sericea* *Alnus rhombifolia / Salix exigua (Rosa californica)* Arroyo Willow (*Salix lasiolepis*) *Baccharis pilularis* / Annual Grasses & Herbs Black Willow (*Salix gooddingii*) Blackberry (*Rubus discolor*) Box Elder (*Acer negundo*) California Wild Rose (*Rosa californica*) Coyotebush (*Baccharis pilularis*) Mexican Elderberry (*Sambucus mexicana*) Narrow-leaf Willow (*Salix exigua*) *Salix exigua -* (*Salix lasiolepis* - Rubus discolor - Rosa californica) *Salix gooddingii / Rubus discolor* *Salix gooddingii /* Wetland Herbs *Salix lasiolepis* - Mixed brambles (*Rosa californica - Vitis californica - Rubus discolor*) Santa Barbara Sedge (*Carex barbarae*) White Alder (*Alnus rhombifolia*) White Alder (*Alnus rhombifolia*) - Arroyo willow (*Salix lasiolepis*) restoration	*Baccharis pilularis* / Annual Grasses & Herbs Blackberry (*Rubus discolor*) Box Elder (*Acer negundo*) Narrow-leaf Willow (*Salix exigua*) *Salix exigua -* (*Salix lasiolepis - Rubus discolor - Rosa californica*) *Salix gooddingii* / wetland herbs *Salix lasiolepis* - Mixed brambles (*Rosa californica - Vitis californica - Rubus discolor*) Southwestern North American introduced riparian scrub Southwestern North American riparian/wash scrub	
Willow thicket	Buttonbush (*Cephalanthus occidentalis*) California Dogwood (*Cornus sericea*) California Hair-grass (*Deschampsia caespitosa*) *Cornus sericea - Salix exigua* *Cornus sericea - Salix lasiolepis / (Phragmites australis)* *Salix lasiolepis - (Cornus sericea) / Scirpus* spp.- (*Phragmites australis - Typha* spp.) complex unit Shining Willow (*Salix lucida*)	Buttonbush (*Cephalanthus occidentalis*)	

Table 5. Table relating modern habitat type map units to non-native/invasive classifications. Values of '1' indicate that the map unit is classified as dominated or co-dominated by non-native or invasive vegetation; values of '0' indicate it is not.

Map unit	Non-native	Inva-sive	Non-native or invasive species
Acacia - Robinia	1	1	Acacia
Acer negundo - Salix gooddingii	0	0	
Agricultural	NA	NA	
Agricultural from 'Agriculture'	NA	NA	
Agricultural from 'Grain/Hay Crops'	NA	NA	
Agriculture	NA	NA	
Algae	0	0	
Alkali Heath (*Frankenia salina*)	0	0	
Alkaline vegetation mapping unit	0	0	
Allenrolfea occidentalis mapping unit	0	0	
Alnus rhombifolia / Cornus sericea	0	0	
Alnus rhombifolia / Salix exigua (Rosa californica)	0	0	
American Bulrush (*Scirpus americanus*)	0	0	
Arid West freshwater emergent marsh	1	0	
Arroyo Willow (*Salix lasiolepis*)	0	0	
Baccharis pilularis / Annual Grasses & Herbs	1	1	Grasslands assumed to be non-native/invasive
Barren	NA	NA	
Black Willow (*Salix gooddingii*)	0	0	
Black Willow (*Salix gooddingii*) - Valley Oak (*Quercus lobata*) restoration	0	0	
Blackberry (*Rubus discolor*)	1	1	*Rubus discolor*
Box Elder (*Acer negundo*)	0	0	
Brazilian Waterweed (*Egeria - Myriophyllum*) Submerged	1	1	*Egeria, Myriophyllum*
Broad-leaf Cattail (*Typha latifolia*)	0	0	
Bromus diandrus - Bromus hordeaceus	1	1	*Bromus diandrus, Bromus hordeaceus*
Buttonbush (*Cephalanthus occidentalis*)	0	0	
California annual forb/grass vegetation	1	1	Grasslands assumed to be non-native/invasive
California Annual Grasslands - Herbaceous	1	1	Grasslands assumed to be non-native/invasive
California Bulrush (*Scirpus californicus*)	0	0	
California Dogwood (*Cornus sericea*)	0	0	
California Hair-grass (*Deschampsia caespitosa*)	0	0	
California Wild Rose (*Rosa californica*)	0	0	
Californian broadleaf forest and woodland	0	0	
Californian mixed annual/perennial freshwater vernal pool/swale/plain bottom-land	0	0	
Californian warm temperate marsh/seep	0	0	
Central and south coastal California seral scrub	0	0	
Coast Live Oak (*Quercus agrifolia*)	0	0	
Common Reed (*Phragmites australis*)	0	0	
Cornus sericea - Salix exigua	0	0	
Cornus sericea - Salix lasiolepis / (Phragmites australis)	0	0	
Coyotebush (*Baccharis pilularis*)	0	0	
Creeping Wild Rye Grass (*Leymus triticoides*)	0	0	
Developed	NA	NA	
Distichlis spicata - Annual Grasses	1	1	Grasslands assumed to be non-native/invasive
Distichlis spicata - Juncus balticus	0	0	

Map unit	Non-native	Inva-sive	Non-native or invasive species
Distichlis spicata - Salicornia virginica	0	0	
Eucalyptus	1	1	*Eucalyptus*
Exotic Vegetation Stands	1	1	Exotic vegetation stands
Floating Primrose (*Ludwigia peploides*)	0	1	*Ludwigia peploides*
Frankenia salina - Distichlis spicata	0	0	
Fremont Cottonwood (*Populus fremontii*)	0	0	
Generic Floating Aquatics	0	0	
Giant Cane (*Arundo donax*)	1	1	*Arundo donax*
Grassland	1	1	Grasslands assumed to be non-native/invasive
Grassland from 'California Annual Grasslands - Herbaceous'	1	1	Grasslands assumed to be non-native/invasive
Grassland from 'Degraded Vernal Pool Complex - California Annual Grasslands - Herbaceous'	1	1	Grasslands assumed to be non-native/invasive
Hard-stem Bulrush (*Scirpus acutus*)	0	0	
Hinds walnut (*Juglans hindsii*)	0	0	
Horsetail (*Equisetum* spp.)	0	0	
Hydrocotyle ranunculoides	0	0	
Intermittently Flooded Perennial Forbs	1	1	*Lepidium latifolium* Semi-natural Stands
Intermittently or Temporarily Flooded Deciduous Shrublands	0	0	
Intermittently or temporarily flooded undifferentiated annual grasses and forbs	1	1	Grasslands assumed to be non-native/invasive
Introduced North American Mediterranean woodland and forest	1	1	Group level: could contain *Eucalyptus, Ailanthus*, and other non-native naturalized trees
Italian Rye-grass (*Lolium multiflorum*)	1	1	*Lolium multiflorum*
Juncus balticus - meadow vegetation	0	0	
Juncus bufonius (salt grasses)	0	0	
Lepidium latifolium - Salicornia virginica - Distichlis spicata	1	1	*Lepidium latifolium*
Levee Rock Riprap	NA	NA	
Lolium multiflorum - Convolvulus arvensis	1	1	*Lolium multiflorum, Convolvulus arvensis*
Lotus scoparius - Antioch Dunes	0	0	
Ludwigia peploides	0	1	*Ludwigia peploides*
Lupinus albifrons - Antioch Dunes	0	0	
Managed alkali wetland (*Crypsis*)	1	0	*Crypsis*
Managed Annual Wetland Vegetation (Non-specific grasses & forbs)	1	1	Undefined, but "likely to be completely dominated by non-natives"
Managed Wetland	NA	NA	
Managed Wetland from 'Agriculture'	NA	NA	
Managed Wetland from 'Rabbitsfoot grass (*Polypogon maritimus*)'	1	0	*Polypogon maritimus*
Mediterranean California naturalized annual and perennial grassland	1	1	Grasslands assumed to be non-native/invasive
Mexican Elderberry (*Sambucus mexicana*)	0	0	
Microphyllous Shrubland	0	0	
Milfoil - Waterweed (generic submerged aquatics)	1	1	Milfoil
Mixed *Scirpus* / Floating Aquatics (*Hydrocotyle - Eichhornia*) Complex	1	1	*Eichhornia, Hydrocotyle*
Mixed *Scirpus* / Submerged Aquatics (*Egeria-Cabomba-Myriophyllum* spp.) complex	1	1	*Egeria, Cabomba, Myriophyllum*
Mixed *Scirpus* Mapping Unit	0	0	
N/A; Agriculture/Non-native/Ruderal	NA	NA	
N/A; Urban/Barren	NA	NA	

Table 5 (continued). Table relating modern habitat type map units to non-native/invasive classifications. Values of '1' indicate that the map unit is classified as dominated or co-dominated by non-native or invasive vegetation; values of '0' indicate it is not.

Map unit	Non-native	Inva-sive	Non-native or invasive species
N/A; Water	NA	NA	
Narrow-leaf Cattail (*Typha angustifolia*)	1	0	*Typha angustifolia*
Narrow-leaf Willow (*Salix exigua*)	0	0	
Narrow-leaf Willow (*Salix exigua*)	0	0	
Naturalized warm-temperate riparian and wetland	NA	NA	
Non-Tidal Perennial Aquatic	NA	NA	
Non-Tidal Perennial Aquatic from 'Agriculture'	NA	NA	
Non-Tidal Perennial Aquatic from 'Water'	NA	NA	
Oregon Ash (*Fraxinus latifolia*)	0	0	
Other Natural Seasonal Wetland	0	0	
Pampas Grass (*Cortaderia selloana - C. jubata*)	1	1	*Cortaderia selloana, Cortaderia jubata*
Perennial Pepperweed (*Lepidium latifolium*)	1	1	*Lepidium latifolium*
Pickleweed (*Salicornia virginica*)	0	0	
Poison Hemlock (*Conium maculatum*)	1	1	*Conium maculatum*
Polygonum amphibium	0	0	
Pondweed (*Potamogeton* sp.)	1	1	*Potamogeton* sp.
Quercus lobata - Acer negundo	0	0	
Quercus lobata - Alnus rhombifolia (Salix lasiolepis - Populus fremontii - Quercus agrifolia)	0	0	
Quercus lobata - Fraxinus latifolia	0	0	
Quercus lobata / Rosa californica (Rubus discolor - Salix lasiolepis / Carex spp.)	1	1	*Rubus discolor, Carex*
Rabbitsfoot grass (*Polypogon maritimus*)	1	0	*Polypogon maritimus*
Restoration Sites	0	0	
Riverine	NA	NA	
Ruderal Herbaceous Grasses & Forbs	1	1	*Silybum marianum, Brassica nigra*
Salicornia virginica - Cotula coronopifolia	1	1	*Cotula coronopifolia*
Salicornia virginica - Distichlis spicata	0	0	
Salix exigua - (Salix lasiolepis - Rubus discolor - Rosa californica)	1	1	*Rubus discolor*
Salix gooddingii - Populus fremontii - (Quercus lobata-Salix exigua-Rubus discolor)	1	1	*Rubus discolor*
Salix gooddingii - Quercus lobata / Wetland Herbs	0	0	
Salix gooddingii / Rubus discolor	1	1	*Rubus discolor*
Salix gooddingii / Wetland Herbs	0	0	
Salix gooddingii / wetland herbs	0	0	
Salix lasiolepis - (Cornus sericea) / Scirpus spp.- *(Phragmites australis - Typha* spp.) complex unit	0	0	
Salix lasiolepis - Mixed brambles (*Rosa californica - Vitis californica - Rubus discolor*)	1	1	*Rubus discolor*
Salt scalds and associated sparse vegetation	0	0	
Saltgrass (*Distichlis spicata*)	0	0	
Santa Barbara Sedge (*Carex barbarae*)	0	0	
Scirpus acutus - (Typha latifolia) - Phragmites australis	0	0	
Scirpus acutus - Typha angustifolia	1	0	*Typha angustifolia*
Scirpus acutus Pure	0	0	
Scirpus acutus -Typha latifolia	0	0	
Scirpus californicus - Eichhornia crassipes	1	1	*Eichhornia crassipes*
Scirpus californicus - Scirpus acutus	0	0	
Scirpus spp. in managed wetlands	0	0	

Map unit	Non-native	Inva-sive	Non-native or invasive species
Seasonally Flooded Grasslands	1	1	Grasslands assumed to be non-native/invasive
Seasonally flooded undifferentiated annual grasses and forbs	1	1	Grasslands assumed to be non-native/invasive
Shallow flooding with minimal vegetation at time of photography	NA	NA	
Shining Willow (*Salix lucida*)	0	0	
Smartweed *Polygonum* spp. - Mixed Forbs	0	0	
Southwestern North American introduced riparian scrub	1	1	Group level: could contain *Arundo donax*, *Tamarix*, *Rubus*
Southwestern North American riparian evergreen and deciduous woodland	0	0	
Southwestern North American riparian/wash scrub	0	0	
Southwestern North American salt basin and high marsh	0	0	
Sparsely or Unvegetated Areas; Abandoned orchards	NA	NA	
Suaeda moquinii - (Lasthenia californica) mapping unit	0	0	
Tall & Medium Upland Grasses	1	1	Grasslands assumed to be non-native/invasive
Temporarily Flooded Grasslands	1	1	*Arundo*; Grasslands assumed to be non-native/invasive
Temporarily Flooded Perennial Forbs	0	0	
Temporarily or Seasonally Flooded - Deciduous Forests	0	0	
Tidal mudflats	NA	NA	
Tidal Perennial Aquatic	NA	NA	
Tidal Perennial Aquatic from 'Brazilian Waterweed (*Egeria - Myriophyllum*) Submerged'	1	1	*Egeria, Myriophyllum*
Tobacco brush (*Nicotiana glauca*) mapping unit	1	1	*Nicotaina glauca*
Tree-of-Heaven (*Ailanthus altissima*)	1	1	*Ailanthis altissima*
Typha angustifolia - Distichlis spicata	1	0	*Typha angustifolia*
Unknown	NA	NA	
Urban	NA	NA	
Urban Developed - Built Up	NA	NA	
Valley Oak (*Quercus lobata*)	0	0	
Valley Oak (*Quercus lobata*) restoration	0	0	
Vernal Pool Complex	0	0	
Vernal Pool Complex from 'California Annual Grasslands - Herbaceous'	1	1	Grasslands assumed to be non-native/invasive
Vernal Pool Complex from 'Vernal Pool - Enhanced'	0	0	
Vernal Pool Complex from 'Vernal Pool - Natural'	0	0	
Vernal Pool Complex from 'Vernal Pools'	0	0	
Vernal Pools	0	0	
Water	NA	NA	
Water Hyacinth (*Eichhornia crassipes*)	1	1	*Eichhornia crassipes*
Western North American Freshwater Aquatic Vegetation	1	1	Group level: could contain *Egeria, Myriophyllum, Ludwigia peploides, Cambomba*, or *Eichhornia crassipes*
White Alder (*Alnus rhombifolia*)	0	0	
White Alder (*Alnus rhombifolia*) - Arroyo willow (*Salix lasiolepis*) restoration	0	0	

Appendix B: Species

The table below lists the common and scientific names of the species mentioned in this report. Bay Delta Conservation Plan Public Draft (BDCP) covered species are marked with an asterisk (*). The Integrated Taxonomic Information System (ITIS) database was used as our nomenclatural reference, except for with names marked with a cross (†), which deviate from those validated by ITIS. The common names of all species are written in lower case, with the following exceptions: (1) the common names of all birds are capitalized, as per American Ornithologists' Union standards and (2) all proper nouns are capitalized. Although the word "Delta" is used as a proper noun throughout this report, we do not capitalize the common name of *Hypomesus transpacificus* (delta smelt), as per U.S. Fish & Wildlife Service standards.

Common name	Scientific name
Birds	
American Wigeon	*Anas americana*
Ash-throated Flycatcher	*Myiarchus cinerascens*
Black Tern	*Chlidonias niger*
Blue Grosbeak	*Passerina caerulea*
Brown-headed Cowbird	*Molothrus ater*
California Black Rail*†	*Laterallus jamaicensis coturniculus*
California Clapper Rail*	*Rallus longirostris obsoletus*
California Least Tern	*Sternula antillarum browni*
Canada Goose	*Branta canadensis*
Canvasback	*Aythya valisineria*
Cinnamon Teal	*Anas cyanoptera*
Common Moorhen	*Gallinula chloropus*
Cooper's Hawk	*Accipiter cooperii*
Downy Woodpecker	*Picoides pubescens*
European Starling	*Sturnus vulgaris*
Forster's Tern	*Sterna forsteri*
Gadwall	*Anas strepera*
Golden-crowned Sparrow	*Zonotrichia atricapilla*
Greater Sandhill Crane*†	*Grus canadensis tabida*
Greater White-fronted Goose	*Anser albifrons*
Green-winged Teal	*Anas crecca*
Hermit Thrush	*Catharus guttatus*
Horned Lark	*Eremophila alpestris*
Least Bell's Vireo*	*Vireo bellii pusillus*
Mallard	*Anas platyrhynchos*
Modesto Song Sparrow†	*Melospiza melodia mailliardi*†
Northern Harrier	*Circus cyaneus*
Northern Pintail	*Anas acuta*
Northern Shoveler	*Anas clypeata*
Oak Titmouse	*Baeolophus inornatus*
Red-shouldered Hawk	*Buteo lineatus*
Ross' Goose	*Chen rossii*

Savannah Sparrow	*Passerculus sandwichensis*
Sharp-shinned Hawk	*Accipiter striatus*
Short-eared Owl	*Asio flammeus*
Snow Goose	*Chen caerulescens*
Suisun Song Sparrow*†	*Melospiza melodia maxillaris*
Swainson's Hawk*	*Buteo swainsoni*
Tricolored Blackbird*	*Agelaius tricolor*
Tundra Swan	*Cygnus columbianus*
Western Burrowing Owl*†	*Athene cunicularia hypugaea*
Western Tanager	*Piranga ludoviciana*
Western Yellow-billed Cuckoo*†	*Coccyzus americanus occidentalis*†
White-tailed Kite*	*Elanus leucurus*
Willow Flycatcher	*Empidonax traillii*
Wilson's Warbler	*Cardellina pusilla*
Wood Duck	*Aix sponsa*
Yellow Warbler	*Setophaga petechia*†
Yellow-breasted Chat*	*Icteria virens*
Yellow Warbler	*Setophaga petechia*
Yellow-rumped Warbler	*Dendroica coronata*†

Fish

bluegill	*Lepomis macrochirus*
California roach	*Hesperoleucus symmetricus*
Chinook salmon*	*Oncorhynchus tshawytscha*
delta smelt*	*Hypomesus transpacificus*
green sturgeon*	*Acipenser medirostris*
hardhead	*Mylopharodon conocephalus*
hitch	*Lavinia exilicauda*
longfin smelt*	*Spirinchus thaleichthys*
Pacific lamprey*	*Entosphenus tridentatus*
Pacific staghorn sculpin	*Leptocottus armatus*
river lamprey*	*Lampetra ayresii*
Sacramento blackfish	*Orthodon microlepidotus*
Sacramento perch	*Archoplites interruptus*
Sacramento pikeminnow	*Ptychocheilus grandis*
Sacramento splittail*	*Pogonichthys macrolepidotus*
Sacramento sucker	*Catostomus occidentalis*
starry flounder	*Platichthys stellatus*
steelhead*	*Oncorhynchus mykiss*
striped bass	*Morone saxatilis*
thicktail chub	*Gila crassicauda*
tule perch	*Hysterocarpus traskii*
white catfish	*Ameiurus catus*
white sturgeon*	*Acipenser transmontanus*

Invertebrates

Asian clam	*Corbicula fluminea*
California linderiella*	*Linderiella occidentalis*
conservancy fairy shrimp*	*Branchinecta conservatio*
Lange's metalmark butterfly	*Apodemia mormo langei*
longhorn fairy shrimp*	*Branchinecta longiantenna*
midvalley fairy shrimp*†	*Branchinecta mesovallensis*
valley elderberry longhorn beetle*	*Desmocerus californicus dimorphus*
vernal pool fairy shrimp*	*Branchinecta lynchi*
vernal pool tadpole shrimp*	*Lepidurus packardi*

Mammals

American beaver	*Castor canadensis*
California ground squirrel	*Otospermophilus beecheyi*
California vole	*Microtus californicus*
coyote	*Canis latrans*
gray fox	*Urocyon cinereoargenteus*
grizzly bear	*Ursus arctos*
long-tailed weasel	*Mustela frenata*
mule deer	*Odocoileus hemionus*
North American river otter	*Lontra canadensis*
pronghorn	*Antilocapra americana*
ringtail	*Bassariscus astutus*
riparian brush rabbit*†	*Sylvilagus bachmani riparius*†
riparian woodrat*†	*Neotoma fuscipes riparia*
salt marsh harvest mouse*	*Reithrodontomys raviventris*
San Joaquin kit fox*†	*Vulpes macrotis mutica*†
San Joaquin Valley kangaroo rat	*Dipodomys nitratoides*
Suisun shrew*†	*Sorex ornatus sinuosus*
tule elk†	*Cervus elaphus nannodes*

Plants

alkali milkvetch*	*Astragalus tener* var. *tener*
Antioch Dunes evening primrose	*Oenothera deltoides* ssp. *howellii*
Boggs Lake hedge-hyssop*	*Gratiola heterosepala*
brittlescale*	*Atriplex parishii* var. *depressa*†
caper-fruited tropidocarpum	*Tropidocarpum capparideum*
Carquinez goldenbush*	*Isocoma arguta*
Delta button celery*†	*Eryngium racemosum*
Delta tule pea*	*Lathyrus jepsonii* var. *jepsonii*
dwarf downingia*†	*Downingia pusilla*
heartscale*	*Atriplex cordulata*
Heckard's peppergrass*†	*Lepidium latipes* var. *heckardii*
Himalayan blackberry†	*Rubus armeniacus*†
iodinebush	*Allenrolfea occidentalis*
legenere*†	*Legenere limosa*†
Mason's lilaeopsis*	*Lilaeopsis masonii*
saltgrass	*Distichlis spicata*

San Joaquin spearscale*†	*Extriplex joaquinana*
side-flowering skullcap*†	*Scutellaria lateriflora*
slough thistle*	*Cirsium crassicaule*
soft bird's-beak*	*Chloropyron molle* ssp. *molle*
Suisun Marsh aster*	*Symphyotrichum lentum*
Suisun thistle*	*Cirsium hydrophilum* var. *hydrophilum*
Welsh mudwort*	*Limosella australis*
western wallflower	*Erysimum capitatum* var. *capitatum*

Reptiles & Amphibians

California red-legged frog*	*Rana draytonii*
California tiger salamander*	*Ambystoma californiense*
giant garter snake*	*Thamnophis gigas*
Western pond turtle*	*Actinemys marmorata*

Endnotes

1 • Introduction

[1] Moyle et al. 2012.

[2] Delta Independent Science Board 2013. Notes that the goals of habitat restoration should emphasize enhancing ecosystem functions and resilience.

[3] Moyle et al. 2012, Cannon and Jennings 2014.

[4] Montgomery 2008. Jackson and Hobbs (2009) note that, "Both our ability to predict where novel ecosystems are heading, and the proactive management of these trajectories, require an understanding of the means by which novel ecosystems develop." The authors continue by stating, "Ecological restoration is rooted in ecological history. To facilitate the recovery of degraded or damaged ecosystems, knowledge of the state of the original ecosystem and what happened to it is invaluable."

[5] Whipple et al. 2012.

[6] Verhoeven et al. (2008) develop the concept of and criteria for determining Operational Landscape Units (OLUs) for restoration visions. This concept was explored for the McCormack-Williamson Tract in the Delta by Beagle et al. (2013), and recommended for further development by Delta Independent Science Board (2013).

[7] Novel ecosystems can be defined as occurring when species are found to exist "in combinations and relative abundances that have not occurred previously within a given biome (Hobbs et al. 2006)," and as the occurrence of assemblages of species that either have not co-occurred historically, or result directly and indirectly from human activities (Bridgewater et al. 2011).

[8] Hanak et al. (2013) report that most people questioned in a widely dispersed survey agreed that discharges of pollutants, direct fish management, changes in the flow regime, invasive species, and alteration of physical habitat have all contributed to the ecosystem decline.

[9] Balaguer et al. 2014.

[10] Atwater and Belknap 1980.

[11] Information on the ecological and physical processes of the historical Delta was gathered and detailed in the *Sacramento-San Joaquin Delta Historical Ecology Investigation* (Whipple et al. 2012)—the source for the summary of the historical Delta landscapes provided in this box.

2 • Project Framework and Methods

[1] Taylor et al. 1990, Brinson 1993, Smith et al. 1995, Jax 2005.

[2] NRC 1995.

[3] Smith et al. 1995.

[4] Hruby et al. 1999.

[5] Delany and Scott 2006.

[6] McGarigal 2002, Kupfer 2012.

[7] McGarigal 2002.

[8] McGarigal 2002.

[9] McGarigal 2002.

[10] Collins and Grossinger 2004.

[11] D'Eon et al. 2002.

[12] Spautz and Nur 2002, Spautz et al. 2005.

[13] Liu et al. 2012.

[14] Spautz and Nur 2002, Spautz et al. 2005.

[15] Gaines 1974.

[16] Kilgo et al. 1998.

[17] Laymon and Halterman 1989.

[18] Whipple et al. 2012.

[19] Hickson and Keeler-Wolf 2007.

[20] GIC 2012.

[21] Daniel Burmester, personal communication; Todd Keeler-Wolf, personal communication.

[22] Whipple et al. 2012.

[23] Whipple et al. 2012.

[24] Gibbes 1850, Ringgold 1850a, Ringgold 1850b.

[25] U.S. Geological Survey 2013. 'NHDArea' layer, high resolution, version 931v210.

[26] Wang and Ateljevich 2012.

[27] Whipple et al. 2012:90.

[28] See Appendix A, pages 96-97 for additional details and specific examples.

[29] Based on work of Hruby et al. (1999).

[30] Based on work of Hruby et al. (1999).

[31] Whipple et al. 2012.

3 • Overall Delta Landscape Changes

[1] Meese et al. 2014.

[2] Yoshiyama et al. 2001.

[3] Garone 2006.

[4] California Department of Water Resources 2013

[5] Mac Nally et al. 2010.

[6] Lund 2010.

[7] Kneib et al. 2008

[8] Whipple et al. 2012.

[9] See Chapter 4 (Life-History Support for Resident and Migratory Fish) for greater detail and references.

[10] Howe and Simenstad 2011.

[11] Couvet 2002, Cushman 2006.

[12] See Chapter 4 (Life-History Support for Resident and Migratory Fish) for greater detail and references.

[13] Lee and Jones-Lee 2004.

[14] Greene et al. 2011.

[15] See Chapter 4 (Life-History Support for Resident and Migratory Fish) for greater detail and references.

[16] Feyrer and Healey 2003.

[17] Atwater and Belknap 1980.

[18] Whipple et al. 2012.

[19] Hickson and Keeler-Wolf 2007.

[20] For example, there are more non-native than native fish species in some parts of the Delta (Feyrer and Healey 2003, Moyle et al. 2012). Species diversity as a restoration goal in the Delta should take into account the role of non-native species.

[21] Modern species-habitat type associations and life-history characteristics were largely derived from BDCP species accounts (California Department of Water Resources 2013), but also from other literature and best professional judgment. Best professional judgment was particularly important for species that today mostly use agricultural lands and managed wetlands. California Department of Water Resources 2013.

[22] Calflora 2013. The Calflora Database, http://www.calflora.org, accessed March 2013.

[23] California Invasive Plant Council (Cal-IPC) 2013. California Invasive Plant Inventory Database, http://www.cal-ipc.org/paf, accessed March 2013.

[24] It is likely that a class of lowest-order tidal channels existed in the Delta that was not represented by historical sources and was thus under-represented in the historical mapping of the Delta (Whipple et al. 2012). We estimate the length of these unmapped channels based on known channel densities in other freshwater marshes in the historical San Francisco Bay-Delta Estuary. See Appendix A for more detail.

[25] Thompson 1957, Enright et al. 2004, Enright 2008.

[26] Modern MLLW elevation was assumed to be 0.64 m NAVD88 (based on data from Cache Slough). Historical MLLW elevation was assumed to be 0.31 m NAVD88. We made the simplifying assumption that the only changes to MLLW since the historical period were from sea level rise (discounting any changes in water surface elevations associated with things like channel armoring, subsidence, and pumping). See Appendix A for additional details.

4 • Life-History Support for Resident and Migratory Fish

[1] Whipple et al. (2012) describe the heterogeneity within aquatic habitats of the historical Delta.

[2] Simenstad et al. (1983). Salmon in the Pacific Northwest used large channels for migration and off-channel habitat for rearing. Smokorowski and Pratt (2007) review how structural habitat complexity supports

diversity in freshwater fish. Features such as undercut banks may be particularly important because of the cover and refuge they provide (e.g., McMahon and Hartman 1989, Cowx and Welcomme 1998).

[3] Whipple et al. (2012) and sources therein.

[4] See Enright (2008) for a discussion of how complex channel networks supported gradients in residence time historically. Enright et al. (2013) explain how channel structure and marsh connection influenced water temperature through geomorphic mediation. Morgan-King and Schoellhamer (2013) describe the processes (e.g., tidal asymmetry) that contribute to the high suspended sediment concentrations observed in the "dead end channels" and "backwaters" of the Cache Slough region.

[5] Sommer et al. (2001a,b), Jeffres et al. (2008), and Opperman (2008) describe the benefits of Delta floodplains, specifically the Cosumnes River and Yolo Bypass, to native fish. Numerous other studies discuss increased prey availability for fish in floodplains in other regions (e.g., Gladden and Smock (1990)).

[6] Hering (2009) details movements of subyearling Chinook salmon to remain in small tidal channels while rearing within the Salmon River Estuary, Oregon. West and Zedler (2000) describe fish use of the marsh plain at high tide, though in a southern California salt marsh. Odum (2000) reviews support for the idea of marshes as productivity sources to estuaries and concludes that the extent of outwelling is related to the extent of marsh, tidal amplitude, and geomorphology, and that large outputs are likely occur as pulses related to storm events and spring tides.

[7] Whipple et al. 2012.

[8] Historical fish assemblages are assumed from modern fish distributions, habitat associations, and life-history requirements. See Moyle (2002) for species-specific information.

[9] Moyle 2002.

[10] Moyle 2002.

[11] Moyle 2002.

[12] Moyle 2002.

[13] Yoshiyama et al. 1998.

[14] Species habitat use is assumed from modern habitat associations and known life-history characteristics as described by Turner (1966), Moyle (2002), Moyle et al. (2004), Crain and Moyle (2011; references from Whipple et al. 2012).

[15] See Whipple et al. (2012:137-142) for discussion of salinity in the historical Delta.

[16] Wiens 2002.

[17] Sommer et al. (2005) discuss fish stranding risk in floodplains, noting that juvenile salmon seek out low-velocity areas on floodplains. The authors also note that although areas with engineered water control structures are associated with comparatively high stranding risk, overall floodplains provide a net benefit to salmon because of the rearing habitat they provide.

[18] See note 14 above.

[19] Hilborne et al. (2003), Greene et al. (2010), and Carlson and Satterthwaite (2011) describe portfolio effects in salmon.

[20] The relationship between residence time and productivity is reviewed in Lucas and Thompson (2012), who describe how the introduction of the invasive overbite clam has altered this relationship.

[21] See notes 4 and 20 above.

[22] Toft et al. 2003.

[23] Howe and Simenstad 2011.

[24] Dependent on allochthonous marsh materials, and likely more so historically. Howe and Simenstad (2011) used stable isotopes to link estuary consumers to primary producer groups in the SF Estuary and found that nearly all sampled organisms relied heavily on allochthonous marsh material. Whitley and Bollens (2014) studied stomach contents of fish at Liberty Island and found tidal marsh was important feeding habitat for many species, including delta smelt, which supplemented their zooplankton-based diet with larval insects in the spring and amphipods in the winter.

[25] Whitley and Bollens (2014) found that prey composition and biomass varied seasonally between fish species at Liberty Island (based on stomach content analysis). Fish maintained stomach fullness with little overlap in diet between species, potentially reducing competition through their flexibility in diet.

[26] Whipple et al. 2012, Opperman 2012.

[27] See note 5 above.

[28] See note 4 above.

[29] Odum (2000) and Kneib et al. (2008) discuss outwelling of organic matter from marshes, though neither discuss the impacts to turbidity directly.

[30] Kneib et al. (2008) and references therein, Howe and Simenstad 2011.

[31] Harmon et al. (1986) and Gregory et al. (1991), among others, review the benefits of large woody debris to anadromous fish. Whipple et al. (2012) describe the location of woody vegetation in the historical Delta.

[32] Sommer et al. 2013.

[33] McIvor and Odum 1988.

[34] Sommer et al. 2001b.

[35] Peter Moyle, personal communication. Also see note 60 below.

[36] Bottom et al. (2005) found that Chinook salmon in the Salmon River Estuary migrated to the ocean over a broader range of sizes and time periods after marsh restoration, suggesting that wetland restoration has expanded life-history variation in the population by allowing greater expression of estuarine-resident behaviors.

[37] See note 39 below. In addition to the loss of environmental cues, Kimmerer (2011) describes the increased risk for passively moving species from water diversions and entrainment.

[38] Bennett et al. (2002) investigated how fish behavior and distribution in multiple species enhanced transport to and retention in nursery habitats in the low salinity zone in the SF Estuary. Fish in this study exhibited behavioral flexibility in different environmental conditions to maximize retention and enhance feeding success. Hering (2009) found salmon moved into and out of tidal marsh channels mostly with the tide, but with some evidence of active movement to enter channels against ebb tides (possibly to maximize foraging efficiency on invertebrate prey exported from the marsh).

[39] Temporally predictable environmental variability can cue reproduction, migration, and other life-history events in Delta fauna (Jassby et al. 1995, Nobriga et al. 2005). See Drinkwater and Frank (1994) for a more general discussion of impacts of river regulation and diversion on fish.

[40] Enright et al. (2013) describes the greater "distance to difference" in modern channel conditions.

[41] Williams et al. 2009. As measured before (1935-1943) and after (1944-1995) the construction of the Shasta Dam.

[42] Whipple et al. 2012.

[43] Kimmerer 2011.

[44] E.g., Werner et al. 2000, Weston and Lydy 2010.

[45] Jassby et al. 2002, Lucas and Thompson 2012, Kimmerer and Thompson 2014.

[46] Toft et al. 2003.

[47] Toft et al. 2003, Brown 2003.

[48] E.g., Moyle 2002, Feyrer and Healey 2003, Nobriga et al. 2005.

[49] Feyrer and Healey (2003) mention striped bass and white catfish as non-natives associated with high flows. Peter Moyle (personal communication) also mentions channel catfish and American shad as additional examples.

[50] Sommer et al. (2013) suggests, however, that invasive species cannot be controlled by changes in hydrology alone.

[51] Peter Moyle, personal communication.

[52] Lab experiments conducted by Marchetti (1999) showed that native Sacramento perch showed reduced growth when placed with non-native bluegill, but only under conditions of food limitation. Peter Moyle (personal communication) notes that food limitation likely also intensifies predation of non-native species on natives. Stephanie Carlson (personal communication) notes that many non-native fish (especially the Centrarchids) are predators. Finally, Sommer et al. (2001a) hypothesize that following flood events the Yolo Bypass becomes a "clean slate" for native fish, who are more adapted to its flood cycle, and thus more able to take advantage of its resources.

[53] Roni et al. (2010) reviewed and modeled coho salmon and steelhead population responses to habitat restoration in Puget Sound and concluded that considerable restoration is needed to produce measurable changes in fish abundance at a watershed scale: "The percentage of floodplain and in-channel habitat that would have to be restored in the modeled watershed to detect a 25% increase in coho salmon and steelhead smolt production (the minimum level detectable by most monitoring programs) was 20%. However, given the large variability in fish response (changes in density or abundance) to restoration, 100% of the habitat would need to be restored to be 95% certain of achieving a 25% increase in smolt production for either species."

[54] Sommer et al. 2001a, Jeffres et al. 2008, Opperman 2008.

[55] Moyle (2002) lists habitat destruction as a possible contributing factor to the decline of Sacramento perch, along with embryo predation and interspecific competition. Thicktail chub "most likely became extinct because they were unable to adapt to the extreme modification of valley floor habitats," and because of the introduction of alien predators.

[56] The timing, frequency, and duration of inundation in the Yolo Bypass is better characterized as 'seasonal short-term flooding' than 'seasonal long-duration flooding.' Historically, water remained on the surface of the Yolo Basin and was available to floodplain-associated fish species for up to six months of the year (it was activated approximately one out of every two years). Since 1944, overflow events into the Bypass of seven days or

longer between mid-March and mid-May occurred in only approximately one out of every four years (Williams et al. 2009). When flooded, the Bypass drains quickly and the extent of inundated habitat varies substantially on the order of days (Sommer et al. 2004).

[57] Sommer et al. (1997) found that strong year classes of splittail, which are obligate floodplain spawners, are not produced unless there are at least three weeks of sustained inundation during the March-April spawning/rearing period. Waples et al. (2009) found that although salmon are equipped with life-history strategies that allow them to persist in disturbance prone environments and across a range of habitats, temporal and spatial access to these ranges of habitats has been limited, resulting in decreased resilience in populations.

[58] Whipple et al. 2012.

[59] This 1927 photograph of North Sacramento shows flooding along the Sacramento River. Photo by McCurry, courtesy of the California History Room, California State Library, Sacramento, California.

[60] Jeffres et al. 2008. This is important because larger juvenile salmon have a higher overall survival rate to adulthood and are more likely to return as spawning adults (Unwin 1997, Galat and Zweimüller 2001). Potential mechanisms for the observed beneficial effects of floodplains include the increased habitat area associated with inundated floodplains (relative to just the adjacent river habitat), which would be expected to reduce resource competition and predator encounter rates (Sommer et al. 2001b), and increase invertebrate prey availability (Gladden and Smock 1990, Sommer et al. 2001b).

[61] This figure represents the combined extent of areas classified as "seasonal short-term flooding" and "seasonal long-duration flooding."

[62] Based on a study from Vogel and Marine (1991) with input from Steve Lindley, personal communication.

[63] Yoshiyama et al. 2001 and Lindley et al. 2004 in Williams 2006:43.

[64] Commissioners of Fisheries 1875, McEwan 2001, and Moyle 2002, as cited in Williams 2006. Williams notes that "the Commissioners of Fisheries (1875:10) also described a summer-run that migrated up the San Joaquin River in July and August that appeared to be '. . . of the same variety as those in the Sacramento, but smaller in size.' The Commission was particularly interested in them because their tolerance of high water temperature '. . . would indicate that they will thrive in all the rivers of the southern states, whose waters take their rise in mountainous or hilly regions. . . .'"

[65] With the exception of spawning, the temporal distributions of the Chinook life-history stages are derived from Vogel and Marine 1991, figure 1, "Life History Characteristics of Sacramento River Chinook Salmon." The noted timing of spawning for each run is taken from Williams 2006 (page 119, table 6-1, "Sacramento River ranges" for fall run; page 120, table 6-2 for late-fall run; page 120, table 6-3 for winter run; and page 121, table 6-4 for spring run). Yoshiyama et al. 2001 and Lindley et al. 2004 in Williams 2006:43.

[66] McEwan 2001:11, figure 3, "Central Valley steelhead life stage periodicity."

[67] Adult migration timing is taken from Moyle et al. 2004, as cited in Kratville 2008:10. The temporal distributions for floodplain/river spawning, embryo and larvae, and juvenile floodplain use are taken from Kratville 2008:3, table 1, "Life stages by biological measures." The listed juvenile downstream migration timing is derived from Moyle et al. 2004, as cited in Kratville 2008:12. Kratville also notes a second life-history strategy for outmigrating juveniles that is not reflected in this table: "a less well studied strategy is to remain upstream through the summer into the next fall or spring and then migrate downstream (Baxter 1999, Moyle et al. 2004). This latter strategy occurs in Butte Creek and the main stem Sacramento River."

[68] Israel and Klimley 2008.

[69] Israel et al. 2009.

[70] USFWS 2012.

[71] Nobriga and Herbold 2009.

[72] Rosenfield 2010.

[73] Gray et al. 2002.

5 • Life-History Support for Marsh Wildlife

[1] Whipple et al. 2012, Vasey et al. 2012.

[2] Described from East Coast marshes by Odum (1988). Moyle et al. (2014) hypothesize that although tidal marshes have lower seed density than managed marshes the extensive acreage of historical marshes in the Bay and Delta would have led to an accumulation of seeds, providing abundant food resources for waterfowl and other wildlife.

[3] Odum 1988.

[4] Reviewed by the DRERIP Tidal Marsh Model and sources therein, Kneib et al. (2008). See Nur et al. (2006) for a description of effects of vegetative structure on the marsh bird community.

[5] Greenberg et al. 2006.

[6] See Herbold and Moyle (1989) and California Department of Water Resources (2013). Mammal species occupying the historical Delta are assumed from distribution of modern native species.

[7] Mitsch and Gosselink (1986) cited in Odum (1988).

[8] Grinnell et al. 1937, Seymour 1960, Gould 1977, Lanman et al. 2013. Description of beavers in the Delta as quoted in Lanman et al. 2013: "There is probably no spot of equal extent in the whole continent of North America which contains so many of these much sought animals (Farnham 1857:383)."

[9] See references in Chapter 4 (Life-History Support for Resident and Migratory Fish).

[10] See Jennings and Hayes (1994) and California Department of Water Resources (2013) for distribution and life-history information on Delta amphibians. These species all require upland habitat for part of their life, which likely prevented them from inhabiting the interior Delta marshes.

[11] See references in Chapter 8 (Life-History Support for Marsh-Terrestrial Transition Zone Wildlife).

[12] Odum 1988.

[13] Based on life-history account of giant garter snake in California Department of Water Resources (2013). Based on life-history account of the Modesto Song Sparrow in Shuford and Gardali (2008). The Modesto Song Sparrow distribution is only slightly broader than the Delta and distinct from the more riparian/upland associated subspecies.

[14] Milliken 1991, Anderson 2005, Manfree 2014 in Moyle et al. 2014.

[15] Whipple et al. 2012.

[16] Lindenmayer and Fischer 2006.

[17] Whigham 1988, Fuji 1998, and Werner et al. 2000.

[18] Lindenmayer and Fischer 2006.

[19] Lindenmayer and Fischer 2006.

[20] CDFG 2007.

6 • Life-History Support for Waterbirds

[1] Historical species occurrences are assumed from modern distributions, life histories and habitat associations. See Herbold and Moyle (1989), California Department of Water Resources (2013), Garone (2011).

[2] Assumptions about waterbird habitat use and ecology were discussed during two meetings with local waterbird experts (Dave Shuford, Daniel Burmester, Dan Skalos, Hildie Spautz, Dave Zezulak) on March 11, 2014 and April 22, 2014. Assumptions of habitat use for particular waterbirds were determined by the best professional judgment of these experts, with the acknowledgement that the magnitude of change in the Delta paired with the large scale at which most waterbirds use the landscape make it difficult to interpret some aspects of waterbird use of the historical Delta. Shorebird habitat associations and the degree to which smaller shorebirds used the Delta were highlighted as areas of particular uncertainty.

[3] Garone 2011.

[4] Central Valley Joint Venture 2006, Whipple et al. 2012.

[5] Garone 2011, Whipple et al. 2012.

[6] Central Valley Joint Venture 2006.

[7] Moyle et al. 2014.

[8] Herbold and Moyle 1989.

[9] See note 2 above.

[10] See note 2 above and Garone 2011.

[11] See note 2 above.

[12] Whipple et al. 2012.

[13] Garone 2011, Whipple et al. 2012.

[14] See note 2 above and Whipple et al. 2012.

[15] See note 2 above.

[16] Ivey et al. 2011, Ivey et al. 2014.

[17] See note 2 above.

[18] See notes 1 and 2 above.

[19] See note 2 above.

[20] Moyle et al. 2014.

[21] Moyle et al. 2014.

[22] Gaines 1980.

[23] Central Valley Joint Venture 2006.

[24] See Miller et al. 2000 and references therein (from Oklahoma). "Agricultural plants are often high in energy, and waterfowl spend more time feeding on crops in the evening to prepare for cold nights. However, feeding exclusively on agricultural crops may not satisfy their protein or mineral requirements. Waterfowl must also include foods that fulfill protein and lipid requirements. Natural plants found in wetlands and invertebrates constitute foods high in protein and amino acids, as well as many minerals."

[25] Mount and Twiss 2005.

[26] See note 2 above.

[27] Moyle et al. 2014.

[28] See notes 1 and 2 above.

[29] See notes 1 and 2 above.

[30] See note 2 above and Garone 2011.

[31] See note 2 above.

[32] Time ranges for wintering and migrating birds are multi-species approximations based on discussions with experts. The breeding waterfowl time range shown is for Mallards.

7 • Life-History Support for Riparian Wildlife

[1] Whipple et al. (2012) describe the position and structure of riparian forests in the historical Delta. The use of riparian forests as movement corridors is well-established (see Hilty and Merenlender (2004) and Fellers and Kleeman (2007) for examples in California).

[2] E.g., Finch 1989.

[3] E.g., Opperman 2002.

[4] Whipple et al. 2012. For Neotropical songbirds, willow-fern marshes may have provided habitat; however, for many less mobile or more terrestrial species, these habitats would have been inaccessible.

[5] See, for example, Brinson et al. (2002) for a discussion of the importance of riparian habitat as a movement corridor for wildlife.

[6] California Department of Water Resources 2013. These species are found primarily in the south Delta today. Whipple et al. (2012) found that the riparian brush rabbit occurred in riparian forests throughout the historical Delta as well.

[7] Gaines 1980 in Sands 1980.

[8] See note 7 above.

[9] See note 7 above.

[10] Geoff Geupel, personal communication.

[11] Thompson (1957) notes that where riparian cover developed historically, "the velocity of sediment laden water was checked," causing natural levees to build up and facilitate more growth of riparian vegetation (a positive feedback cycle).

[12] California Department of Water Resources 2013.

[13] Gaines 1980.

[14] Small 2005, Small et al. 2007, Golet et al. 2008.

[15] Whisson et al. 2007, California Department of Water Resources 2013.

[16] California Department of Water Resources 2013.

[17] Seavy et al. 2009.

[18] Laymon and Halterman 1989.

[19] Measured as the maximum geodesic distance (as the crow flies) an organism can travel away from a starting location within a single contiguous woody riparian habitat polygon (defined by the minimum mapping unit).

[20] The 100 m threshold for grouping riparian polygons into patches is based on the typical maximum gap crossing distance of dispersing songbirds, as determined by best professional judgment (Geoff Geupel, personal communication).

[21] Fischer 2000.

[22] Whipple et al. (2012) mapped the dominant habitat types, so while the Cosumnes area appears to be absent of woody riparian vegetation, there were likely some wooded sloughs and willow thickets that were too small to map.

8 • Life-History Support for Marsh-Terrestrial Transition Zone Wildlife

[1] Whipple et al. 2012, "Pattern of edge."

[2] Assumed from life-history characteristics. See BDCP species accounts (California Department of Water Resources 2013) and references therein.

[3] See Chapter 7 (Life-History Support for Riparian Wildlife) for greater detail and references.

[4] Whipple et al. 2012, "Tule elk breeding on dunes."

[5] Whipple et al. 2012, "Variable seasonal wetlands."

[6] California Department of Water Resources 2013. Species of Concern.

[7] See Chapter 6 (Life-History Support for Waterbirds) and California Department of Water Resources 2013.

[8] California Department of Water Resources 2013.

[9] Trapp 2011.

[10] Barbour et al. 2007.

[11] California Department of Water Resources 2013.

[12] Whipple et al. 2012.

[13] California Department of Water Resources 2013.

[14] Whipple et al. 2012, Schiffman 2011.

[15] Schiffman 2011, Trapp 2011, California Department of Water Resources 2013.

[16] Trapp 2011.

[17] Milliken 1991, Anderson 2005, Manfree 2014 in Moyle et al. 2014.

[18] Goals Project 2014 (in development).

Appendix A: Methods

[1] Whipple et al. 2012.

[2] Whipple et al. 2012.

[3] Whipple et al. 2012, available for download at http://www.sfei.org/sites/default/files/Delta_Historical_ Ecology_GISdata_SFEI_ASC_2012.zip.

[4] Hickson and Keeler-Wolf 2007.

[5] GIC 2012.

[6] WWR 2013.

[7] California Department of Water Resources 2013.

[8] Whipple et al. 2012.

[9] Daniel Burmester and Todd Keeler-Wolf, personal communication.

[10] Whipple et al. 2012.

[11] **Buck-Diaz et al. 2012.**

[12] Daniel Burmester and Todd Keeler-Wolf, personal communication.

[13] CALFED 2000.

[14] Todd Keeler-Wolf, personal communication.

[15] Whipple et al. 2012.

[16] Hickson and Keeler-Wolf 2007.

[17] See California Department of Water Resources 2013, appendix 2.b for detailed methodology.

[18] See note 3 above.

[19] See note 3 above.

[20] U.S. Geological Survey 2013. 'NHDArea' layer, high resolution, version 931v210.

[21] CWS et al. 2014. Manuscript in preparation.

[22] Cordell 1867.

[23] California Debris Commission (Debris Commission) 1908-1913. Since the Debris Commission surveys took place after substantial alteration of Delta waterways from hydraulic mining debris, channel cuts, and dredging, we limited our use of Debris Commission bathymetric data to channel reaches with minimal apparent physical alteration.

[24] The maps produced by Ringgold (1850a & 1850b) and Gibbes (1850) lack the spatial accuracy of the USCS hydrographic sheet and have no known projection or features from which to establish reliable control points. We were thus unable to directly digitize historical soundings from georeferenced maps. The soundings recorded by Ringgold and Gibbes were instead georeferenced by matching channel meanders

and confluences on the historical maps with meanders and confluences in the Delta Historical Ecology channel centerline layer (soundings were generally taken at the apex of meanders) and placing the soundings relative to these features. Any soundings that were difficult to place were discarded.

[25] The parabolic channel shape was chosen after conversations with experts on tidal channel morphology. While this shape inevitably simplifies channel morphology, we felt it best represented channel cross-sectional area given the available data. CWS technicians applied the parabolic shape by calculating parabolic channel cross-sections between the historical channel thalweg and shoreline (set to a depth of 0 m/MLLW) at 100 m intervals and outputting these cross-sections as a series of points. These points were converted to modern fixed datum (NAVD88, see Appendix A, Section 4.1.4) and then used as TIN inputs to generate continuous DEM bathymetry.

[26] Atwater et al. 1977.

[27] Whipple et al. 2012.

[28] Wang and Ateljevich 2012, version 3.

[29] Wang and Ateljevich 2012.

[30] cbec 2010.

[31] Whipple et al. 2012.

[32] Phil Williams, personal communication.

[33] Grossinger 1995.

[34] Collins and Grossinger 2004.

[35] Whipple et al. 2012.

[36] cbec 2010.

[37] See Lopez et al. 2006.

[38] Simenstad 1983.

[39] Ashley and Zeff 1988.

[40] Simenstad 1983, Collins 1998.

[41] Ashley and Zeff 1988.

[42] Morgan-King and Schoellhammer 2013.

[43] E.g., Pethick 1992.

[44] Simenstad 1983, Collins 1998, Hood 2006.

[45] Cavallo et al. 2013.

[46] Whipple et al. 2012:331-333.

[47] Whipple et al. 2012:38.

[48] Alison Whipple, personal communication. Dense tidal channel networks served as an indicator of daily tidal inundation, especially in the lower/southern portion of the Yolo Basin tidal area. Historical quotes about tides flowing in and out of lower Grand, Staten, and Tyler islands increased confidence that the Cache Slough region experienced daily tidal inundation.

[49] Whipple et al. 2012:127-128.

[50] Rose et al. 1895 in Whipple et al. 2012:128.

[51] Most information on depth, duration, and timing is derived from Whipple et al. (2012). Additional information on the depth of historical inundation was obtained from historical General Land Office surveys of the Delta.

[52] Pasternak et al. 2004.

[53] D'Eon et al. 2002.

[54] Spautz and Nur 2002, Spautz et al. 2005.

[55] Liu et al. 2012.

[56] Spautz and Nur 2002, Spautz et al. 2005.

[57] Spautz and Nur 2002, Spautz et al. 2005.

[58] Geoffrey Geupel, personal communication.

[59] D'Eon et al. 2002.

[60] Laymon and Halterman 1989.

[61] Whipple et al. 2012:178-183.

[62] Gibbes 1850. See Whipple et al. 2012, figure 4.49.

[63] From Whipple et al. 2012, figure 4.50.

[64] DWR 2014. California Levee Database Centerlines, Version 3, Release 2.

[65] Whipple et al. 2012.

[66] Hickson and Keeler-Wolf 2007.

[67] CDFG 2011.

[68] Hickson and Keeler-Wolf 2007.

[69] Daniel Burmester, personal communication.

References

Anderson K. 2005. *Tending the wild: Native American knowledge and the management of California's natural resources*. Berkeley, CA: University of California Press.

Ashley GM, Zeff ML. 1988. Tidal channel classification for a low-mesotidal salt marsh. *Marine Geology* 82(1):17-32.

Atwater BF, Belknap DF. 1980. Tidal-wetland deposits of the Sacramento-San Joaquin Delta, California. In *Quaternary depositional environments of the Pacific Coast: Pacific Coast Paleogeography Symposium 4*, ed. M.E. Field, A.H. Bouma, I.P. Colburn, R.G. Douglas, and J.C. Ingle. Los Angeles, California: The Pacific Section Society of Economic Paleontologists and Mineralogists.

Atwater BF, Hedel CW, Helley EJ. 1977. *Late Quaternary depositional history, Holocene sea-level changes, and vertical crust movement, southern San Francisco Bay, California*. US Govt. Print. Off.

Aquatic Science Center. 2011. *The Pulse of the Delta: Monitoring and Managing Water Quality in the Sacramento–San Joaquin Delta*. Aquatic Science Center, Oakland, CA.

Balaguer L, Escudero A, Martín-Duque JF, Mola I, Aronson J. 2014. The historical reference in restoration ecology: re-defining a cornerstone concept. *Biological Conservation* 176:12-20.

Barbour MG, Keeler-Wolf T, Schoenherr AA. 2007. *Terrestrial vegetation of California*. Univ of California Press.

Baxter RD. 1999. *Splittail abundance and distribution update*. Available at: http://www2.delta.a.gov/reports/splittail/abundance/html

Beagle JR, Whipple AA, Grossinger RM. 2013. *Landscape patterns and processes of the McCormack-Williamson Tract and surrounding area: A framework for restoring a resilient and functional landscape*. San Francisco Estuary Institute-Aquatic Science Center, Richmond, CA.

Bennett WA, Kimmerer WJ, Burau JR. 2002. Plasticity in vertical migration by native and exotic estuarine fishes in a dynamic low-salinity zone. *Limnology and Oceanography* 47(5):1496-1507.

Bottom DL, Jones KK, Cornwell TJ, Gray A, Simenstad CA. 2005. Patterns of Chinook salmon migration and residency in the Salmon River estuary (Oregon). *Estuarine, Coastal and Shelf Science* 64(1):79-93.

Bridgewater P, Higgs E, Hobbs R, Jackson S. 2011. Engaging with novel ecosystems. *Frontiers in Ecology and the Environment* 9.

Brinson M, MacDonnell L, Austen D, Beschta R, Dillaha T, Donahue D, Gregory S, Harvey J, Molles M, Rogers E. 2002. Riparian areas: functions and strategies for management. *National Academy of Sciences, Washington DC*.

Brinson MM. 1993. *A hydrogeomorphic classification for wetlands*. U.S. Army Corps of Engineers.

Brown LR. 2003. Will tidal wetland restoration enhance populations of native fishes? *San Francisco Estuary and Watershed Science* 1(1).

Buck-Diaz J, Batiuk S, Evens JM. 2012. Vegetation alliances and associations of the Great Valley Ecoregion, California. *California Native Plant Society, Sacramento. Final report to the Geographical Information Center, Chico State University*.

CALFED Bay-Delta Program. 2000. *Multi-Species Conservation Strategy: Final Programmatic EIS/EIR Technical Appendix*. CALFED Bay-Delta Program.

Calflora. 2013. www.calflora.org.

California Debris Commission (Debris Commission). 1908-1913. Maps of the Sacramento, Feather, American, Mokelumne, and San Joaquin Rivers. San Francisco, CA.

California Department of Fish and Game (CDFG). 2007. Lower Sherman Island Wildlife Area Land Management Plan. Rancho Cordova, California. April 2007.

California Department of Fish and Game (CDFG). 2011. *Accuracy Assessment of Mid-Scale Central Valley Riparian Vegetation Map*. Developed for California Department of Water Resources Central Valley Flood Protection Program and Geographical Information Center California State University, Chico.

California Department of Water Resources (DWR). 2013. *Bay Delta Conservation Plan*. Prepared by ICF International Sacramento, CA.

California Department of Water Resources (DWR). 2014. California Levee Database. Available: http://www.water.ca.gov/floodmgmt/lrafmo/fmb/fes/levee_database.cfm

California Water Code 85302 (e)(1). Available: http://www.leginfo.ca.gov/.html/wat_table_of_contents.html

Cannon T, Jennings B. 2014. An overview of habitat restoration successes and failures in the Sacramento-San Joaquin Delta. California Sportfishing Protection Alliance.

Carlson SM, Satterthwaite WH. 2011. Weakened portfolio effect in a collapsed salmon population complex. *Canadian Journal of Fisheries and Aquatic Sciences* 68(9):1579-1589.

Cavallo B, Merz J, Setka J. 2013. Effects of predator and flow manipulation on Chinook salmon (*Oncorhynchus tshawytscha*) survival in an imperiled estuary. *Environmental Biology of Fishes* 96(2-3):393-403.

cbec. 2010. BDCP Effect Analysis: 2D Hydrodynamic Modeling of the Fremont Weir Diversion Structure. Prepared for SAIC and the California DWR. November, 2010.

Center for Watershed Sciences (CWS), San Francisco Estuary Institute (SFEI), Resource Management Associates (RMA). 2014. Developing a continuous topographic-bathymetric digital elevation model (DEM) of the historical (ca. 1800) Sacramento-San Joaquin Delta. *Manuscript in preparation.*

Central Valley Joint Venture. 2006. Central Valley Joint Venture implementation plan-conserving bird habitat. US Fish and Wildlife Service Sacramento, California.

Collins B. 1998. Preliminary assessment of historic conditions of the Skagit River in the Fir Island area: Implications for salmonid habitat restoration. *Skagit System Cooperative, La Conner, WA*.

Collins JN, Grossinger RM. 2004. *Synthesis of scientific knowledge concerning estuarine landscapes and related habitats of the South Bay Ecosystem*. San Francisco Estuary Institute, Oakland.

Commissioners of Fisheries of the State of California. 1875. *Report for the years 1874 and 1875*. Sacramento, CA.

Cordell E. 1867. Hydrography of part of Sacramento and San Joaquin Rivers, California. U.S. Coast Survey (USCS). Register No. 935. 1:10,000.

Couvet D. 2002. Deleterious effects of restricted gene flow in fragmented populations. *Conservation Biology* 16(2):369-376.

Cowx IG, Welcomme RL. 1998. *Rehabilitation of Rivers for Fish*. Fishing News Books, Oxford, UK.

Crain PK, Moyle PB. 2011. Biology, history, status, and conservation of the Sacramento perch, *Archoplites interruptus*: a review. *San Francisco Estuary and Watershed Science* 9(1):1-35.

Cushman SA. 2006. Effects of habitat loss and fragmentation on amphibians: a review and prospectus. *Biological Conservation* 128(2):231-240.

Delany S, Scott D. 2006. *Waterbird Population Estimates – Fourth Edition*. Wetlands International, Wageningen, The Netherlands.

Delta Independent Science Board. (2013). Habitat Restoration in the Sacramento –San Joaquin Delta and Suisun Marsh: A review of Science Programs. Delta Stewardship Council.

Delta Stewardship Council. (2013). The Delta Plan: Ensuring a reliable water supply for California, a healthy Delta ecosystem, and a place of enduring value.

D'Eon R, Glenn SM, Parfitt I, Fortin M-J. 2002. Landscape connectivity as a function of scale and organism vagility in a real forested landscape. *Conservation Ecology* 6(2):10.

Drinkwater KF, Frank KT. 1994. Effects of river regulation and diversion on marine fish and invertebrates. *Aquatic Conservation: Marine and Freshwater Ecosystems* 4(2):135-151.

Enright C. 2008. *Tidal Slough "Geometry" Filters Estuarine Drivers, Mediates Transport Processes, and Controls Variability of Ecosystem Gradients*. Calfed Science Conference proceeding, Sacramento CA.

Enright C, Culberson SD, Burau JR. 2013. Broad Timescale Forcing and Geomorphic Mediation of Tidal Marsh Flow and Temperature Dynamics. *Estuaries and Coasts* 36(6):1319-1339.

Enright CS, Miller A, Tom B. 2004. The Estuary Geometry Is Not Static: Perspective on Salinity Trends From Natural and Human Influences. CALFED Science Conference 2004, Sacramento, CA.

Fellers GM, Kleeman PM. 2007. California red-legged frog (*Rana draytonii*) movement and habitat use: implications for conservation. *Journal of Herpetology* 41(2):276-286.

Feyrer F, Healey MP. 2003. Fish community structure and environmental correlates in the highly altered southern Sacramento-San Joaquin Delta. *Environmental Biology of Fishes* 66(2):123-132.

Finch DM. 1989. Habitat use and habitat overlap of riparian birds in three elevational zones. *Ecology*:866-880.

Fischer R, Theriot RF. 2000. Width of Riparian Zones for Birds. Army Engineer Waterways Experiment Station Vicksburg MA Engineer Research and Development Center, 2000.

Gaines D. 1974. Review of the status of the Yellow-billed Cuckoo in California: Sacramento Valley populations. *Condor*:204-209.

Gaines DA. 1980. *The valley riparian forests of California: their importance to bird populations*. Riparian Forests in California: Their Ecology and Conservation: a Symposium, 57.

Galat DL, Zweimüller I. 2001. Conserving large-river fishes: is the highway analogy an appropriate paradigm? *Journal of the North American Benthological Society* 20(2):266-279.

Garone PF. 2006. The Fall and Rise of the Wetlands of California's Great Central Valley: A Historical and Ecological Study of an Endangered Resource of the Pacific Flyway. Doctor of Philosophy, History, University of California, Davis.

Garone PF. 2011. *The Fall and Rise of the Wetlands of California's Great Central Valley*. Berkeley and Los Angeles: University of California Press.

Geographical Information Center CSU, Chico (GIC). 2012. *Mapping Standard and Land Use Categories for the Central Valley Riparian Mapping Project*. Developed for the Central Valley Flood Protection Program Systemwide Planning Area, major rivers and tributaries.

Gibbes CD. 1850. Map of San Joaquin River. San Francisco, CA: W.B. Cooke & Co. *Courtesy of Peter J. Shields Library Map Collection, UC Davis*.

Gladden JE, Smock LA. 1990. Macroinvertebrate distribution and production on the floodplains of two lowland headwater streams. *Freshwater Biology* 24(3):533-545.

Goals Project. 2014 (in development). *The Baylands and Climate Change: What We Can Do. The 2014 Science Update to the Baylands Ecosystem Habitat Goals prepared by the San Francisco Bay Area Wetlands Ecosystem Goals Project*. California State Coastal Conservancy, Oakland, CA.

Golet GH, Gardali T, Howell CA, Hunt J, Luster RA, Rainey W, Roberts MD, Silveira J, Swagerty H, Williams N. 2008. Wildlife response to riparian restoration on the Sacramento River. *San Francisco Estuary and Watershed Science* 6(2).

Gould GIJ. 1977. *Status of the River Otter in California*. Department of Fish and Game.

Gray A, Simenstad CA, Bottom DL, Cornwell TJ. 2002. Performance of Juvenile Salmon Habitat in Recovering Wetlands of the Salmon River Estuary, Oregon, U.S.A. *Restoration Ecology* 10(3):514–526.

Greenberg R, Maldonado JE, Droege S, McDonald M. 2006. Tidal marshes: a global perspective on the evolution and conservation of their terrestrial vertebrates. *BioScience* 56(8):675-685.

Greene CM, Hall JE, Guilbault KR, Quinn TP. 2010. Improved viability of populations with diverse life-history portfolios. *Biology Letters* 6(3):382-386.

Greene VE, Sullivan LJ, Thompson JK, Kimmerer WJ. 2011. Grazing impact of the invasive clam Corbula amurensis on the microplankton assemblage of the northern San Francisco Estuary. *Marine Ecology Progress Series* 431:183-193.

Gregory SV, Swanson FJ, McKee WA, Cummins KW. 1991. An ecosystem perspective of riparian zones. *BioScience*:540-551.

Grinnell J, Dixon JS, Linsdale JM. 1937. *Fur-bearing mammals of California: their natural history, systematic status, and relations to man*. Berkeley, CA: University of California Press.

Grossinger RM. 1995. *Historical evidence of freshwater effects on the plan form of tidal marshlands in the Golden Gate Estuary*. University of California, Santa Cruz.

Hall WH. ca. 1880. Grand Island and Suisun Bay to foothills and 1st Standard North. *Courtesy of California State Archives, Sacramento*.

Hanak E, Phillips C, Lund J, Durand J, Mount J, Moyle P. 2013. Scientist and Stakeholder Views on the Delta Ecosystem. *San Francisco, CA: Public Policy Institute of California*.

Harmon ME, Franklin JF, Swanson FJ, Sollins P, Gregory S, Lattin J, Anderson N, Cline S, Aumen N, Sedell J. 1986. Ecology of coarse woody debris in temperate ecosystems. *Advances in Ecological Research* 15(133):302.

Herbold B, Moyle PB. 1989. *The ecology of the Sacramento-San Joaquin Delta: a community profile*. U.S. Fish and Wildlife Service, Washington, D.C.

Hering DK. 2009. Growth, residence, and movement of juvenile Chinook salmon within restored and reference estuarine marsh channels in Salmon River, Oregon. Master's thesis, Oregon State University.

Hickson D, Keeler-Wolf T. 2007. *Vegetation and land use classification and map of the Sacramento-San Joaquin River Delta*. Report prepared for Bay Delta Region of California Department of Fish and Game, Sacramento, CA.

Hilborn R, Quinn TP, Schindler DE, Rogers DE. 2003. Biocomplexity and fisheries sustainability. *Proceedings of the National Academy of Sciences* 100(11):6564-6568.

Hilty JA, Merenlender AM. 2004. Use of riparian corridors and vineyards by mammalian predators in northern California. *Conservation Biology* 18(1):126-135.

Hobbs RJ, Arico S, Aronson J, Baron JS, Bridgewater P, Cramer VA, Epstein PR, Ewel JJ, Klink CA, Lugo AE, Norton D, Ojima D, Richardson DM, Sanderson EW, Valladares F, Vilà M, Zamora R, Zobel M. 2006. Novel ecosystems: theoretical and management aspects of the new ecological world order. *Global Ecology and Biogeography* 15:1-7.

Hood WG. 2006. A conceptual model of depositional, rather than erosional, tidal channel development in the rapidly prograding Skagit River Delta (Washington, USA). *Earth Surface Processes and Landforms* 31(14):1824-1838.

Howe ER, Simenstad CA. 2011. Isotopic determination of food web origins in restoring and ancient estuarine wetlands of the San Francisco Bay and Delta. *Estuaries and Coasts* 34(3):597-617.

Hruby T, Granger T, Brunner K, Cooke S, Dublonica K, Gersib R, Granger T, Reinelt L, Richter K, Sheldon D. 1999. Methods for assessing wetland functions. *Volume I: Riverine and Depressional Wetlands in the Lowlands of Western Washington. Ecology Publication*:99-115.

Israel J, Drauch A, Gingras M. 2009. *Life History Conceptual Model for White Sturgeon (Acipenser transmontanus)*. University of California, Davis, Stockton.

Israel JA, Klimley AP. 2008. *Life History Conceptual Model for North American Green Sturgeon (Acipenser medirostris)*. University of California, Davis.

Ivey GL, Dugger BD, Herziger CP, Casazza ML, Fleskes JP. 2011. *Sandhill Crane Use of Agricultural Landscapes in the Sacramento-San Joaquin Delta Region of California*. US Geological Survey.

Ivey GL, Dugger BD, Herziger CP, Casazza ML , Fleskes JP. 2014. Characteristics of Sandhill Crane roosts in the Sacramento-San Joaquin Delta of California. *Proc. North Am. Crane Workshop*.

Jackson S, Hobbs R. 2009. Ecological restoration in the light of ecological history. *Science* 325:567-569.

Jassby A. 2008. Phytoplankton in the Upper San Francisco Estuary: Recent biomass trends, their causes and their trophic significance. *San Francisco Estuary and Watershed Science* 6(1).

Jassby AD, Cloern JE, Cole BE. 2002. Annual primary production: patterns and mechanisms of change in a nutrient-rich tidal ecosystem. *Limnology and Oceanography* 47(3):698-712.

Jassby AD, Kimmerer WJ, Monismith SG, Armor C, Cloern JE, Powell TM, Schubel JR, Vendlinski TJ. 1995. Isohaline Position as a Habitat Indicator for Estuarine Populations. *Ecological Applications* 5(1):1995.

Jax K. 2005. Function and "functioning" in ecology: what does it mean? *Oikos* 111(3):641-648.

Jeffres CA, Opperman JJ, Moyle PB. 2008. Ephemeral floodplain habitats provide best growth conditions for juvenile Chinook salmon in a California river. *Environmental Biology of Fishes* 83(4):449-458.

Jennings MR, Hayes MP. 1994. *Amphibian and Reptile Species of Special Concern in California*. California Department of Fish and Game, Rancho Cordova.

Kilgo JC, Sargent RA, Chapman BR, Miller KV. 1998. Effect of stand width and adjacent habitat on breeding bird communities in bottomland hardwoods. *The Journal of wildlife management*:72-83.

Kimmerer WJ. 2011. Modeling Delta Smelt losses at the south Delta export facilities. *San Francisco Estuary and Watershed Science* 9(1).

Kimmerer WJ, Thompson JK. 2014. Phytoplankton growth balanced by clam and zooplankton grazing and net transport into the low-salinity zone of the San Francisco Estuary. *Estuaries and Coasts*:1-17.

Kneib RT, Simenstad CA, Nobriga ML, Talley DM. 2008. *Tidal marsh conceptual model*. Sacramento (CA): Delta Regional Ecosystem Restoration Implementation Plan.

Kratville D. 2008 *Sacramento Splittail Pogonichthys macrolepidotus*. Department of Fish and Game.

Kupfer JA. 2012. Landscape ecology and biogeography: Rethinking landscape metrics in a post-FRAGSTATS landscape. *Progress in Physical Geography*:0309133312439594.

Lanman CW, Lundquist K, Perryman H, Asarian JE, Dolman B, Lanman RB, Pollock MM. 2013. The historical range of beaver (*Castor canadensis*) in coastal California: an updated view of the evidence. *California Fish and Game* 99(4):193-221.

Laymon SA, Halterman MD. 1989. A proposed habitat management plan for yellow-billed cuckoos in California. *Forest Service General Technical Report PSW-110*:272-277.

Lee GF, Jones-Lee A. 2004. *Overview of Sacramento-San Joaquin River Delta Water Quality Issues*. G. Fred Lee & Associates.

Lindenmayer DB, Fischer J. 2006. *Habitat fragmentation and landscape change: an ecological and conservation synthesis*. Washington: Island Press.

Lindley ST, Schick R, May B, Anderson J, Greene S, Hanson C, Low A, McEwan D, MacFarlane R, Swanson C. 2004. *Population structure of threatened and endangered Chinook salmon ESUs in California's Central Valley basin*. US Department of Commerce, National Oceanic and Atmospheric Administration, National Marine Fisheries Service, Southwest Fisheries Science Center.

Liu L, Wood J, Nur N, Salas L, Jongsomjit D. 2012. *California Clapper Rail (Rallus longirostris obsoletus) Population monitoring: 2005-2011*. PRBO Conservation Science, Petaluma, CA.

Lopez CB, Cloern JE, Schraga TS, Little AJ, Lucas LV, Thompson JK, Burau JR. 2006. Ecological values of shallow-water habitats: Implications for the restoration of disturbed ecosystems. *Ecosystems* 9(3):422-440.

Lucas LV, Thompson JK. 2012. Changing restoration rules: Exotic bivalves interact with residence time and depth to control phytoplankton productivity. *Ecosphere* 3(12):art117.

Lund JR. 2010. *Comparing futures for the Sacramento-San Joaquin delta*. Univ of California Press.

Mac Nally R, Thomson JR, Kimmerer WJ, Feyrer F, Newman KB, Sih A, Bennett WA, Brown L, Fleishman E, Culberson SD, Castillo G. 2010. Analysis of pelagic species decline in the upper San Francisco Estuary using multivariate autoregressive modeling (MAR). *Ecological Applications* 20(5):1417-1430.

Manfree AD. 2014. Historical ecology. In Moyle PB, Manfree AD, Fiedler PL. 2014. *Suisun Marsh: Ecological History and Possible Futures*. University of California Press.

Marchetti MP. 1999. An experimental study of competition between the native Sacramento perch (*Archoplites interruptus*) and introduced bluegill (*Lepomis macrochirus*). *Biological Invasions* 1(1):55-65.

McEwan D. 2001. Central Valley steelhead. In *Contributions to the biology of Central Valley salmonids*, ed. R L Brown: California Department of Fish and Game, Fish Bulletin 179.

McGarigal K. 2002. Landscape pattern metrics. Encyclopedia of environmetrics.

McIvor CC, Odum WE. 1988. Food, predation risk, and microhabitat selection in a marsh fish assemblage. *Ecology*:1341-1351.

McMahon TE, Hartman GF. 1989. Influence of cover complexity and current velocity on winter habitat use by juvenile coho salmon (*Oncorhynchus kisutch*). *Canadian Journal of Fisheries and Aquatic Sciences* 46(9):1551-1557.

Meese RJ, Beedy EC, Hamilton WJ. 2014. Tricolored Blackbird (*Agelaius tricolor*). The Birds of North America Online (A. Poole, Ed.). Cornell Lab of Ornithology, Ithaca.

Miller OD, Wilson JA, Ditchkoff SS, Lochmiller RL. 2000. Consumption of Agricultural and Natural Foods by Waterfowl Migrating Through Central Oklahoma. *Proceedings of the Oklahoma Academy of Science* 80:25-31.

Milliken R. 1991. An ethnohistory of the Indian people of the San Francisco Bay Area from 1770 to 1810. University of California, Berkeley.

Mitsch WJ, Gosselink JG. 1986. *Wetlands*. New York: Van Nostrand Reinhold.

Montgomery DR. 2008. Dreams of natural streams. *Science* 319(5861):291-292.

Morgan-King TL, Schoellhamer DH. 2013. Suspended-Sediment Flux and Retention in a Backwater Tidal Slough Complex near the Landward Boundary of an Estuary. *Estuaries and Coasts* 36(2):300-318.

Mount J, Twiss R. 2005. Subsidence, Sea Level Rise, and Seismicity in the Sacramento-San Joaquin Delta. *San Francisco Estuary and Watershed Science* 3(1).

Moyle P, Bennett W, Durand J, Fleenor W, Bray B, Hanak E, Lund J, Mount J. 2012. *Where the wild things aren't: Reconciling the Sacramento-San Joaquin Delta Ecosystem.* Public Policy Institute of California.

Moyle PB. 2002 *Inland fishes of California. Revised and expanded.* Berkeley, CA: University of California Press.

Moyle PB, Baxter RD, Sommer T, Foin TC, Matern SA. 2004. Biology and Population Dynamics of Sacramento Splittail in the San Francisco Estuary: A Review. *San Francisco Estuary and Watershed Science* 2(2).

Moyle PB, Manfree AD, Fiedler PL. 2014. *Suisun Marsh: Ecological History and Possible Futures.* University of California Press.

National Agriculture Imagery Program (NAIP). 2005. [Natural color aerial photos of Contra Costa, Sacramento, San Joaquin, Solano, Yolo counties]. Ground resolution: 1m. U.S. Department of Agriculture (USDA). Washington, DC.

National Research Council. 1995. Wetlands: characteristics and boundaries. National Research Council, National Academy of Sciences. National Academy Press, Washington, p. 328.

Nobriga ML, Feyrer F, Baxter RD, Chotkowski M. 2005. Fish community ecology in an altered river delta: spatial patterns in species composition, life history strategies, and biomass. *Estuaries* 28(5):776-785.

Nobrigia M, Herbold B. 2009. *The Little Fish in California's Water Supply: a Literature Review and Life-History Conceptual Model for delta smelt (Hypomesus transpacificus) for the Delta Regional Ecosystem Restoration and Implementation Plan (DRERIP).* California Department of Fish and Game.

Nur N, Herzog M, Liu L, Kelly J, Evans J, Stralberg D, Warnock N. 2006. *Integrated Regional Wetland Monitoring Pilot Project Bird Team Data Report.* PRBO Conservation Science, Petaluma.

Odum EP. 2000. Tidal marshes as outwelling/pulsing systems. In *Concepts and controversies in tidal marsh ecology*, ed., 3-7: Springer.

Odum WE. 1988. Comparative Ecology of Tidal Freshwater and Salt Marshes. *Annual Review of Ecology and Systematics* 19:147-176.

Opperman JJ. 2002. Anadromous fish habitat in California's Mediterranean-climate watersheds: influences of riparian vegetation, instream large woody debris, and watershed-scale land use. University of California, Berkeley.

Opperman JJ. 2008. *Floodplain ecosystem conceptual model.* Sacramento (CA): Delta Regional Ecosystem Restoration Implementation Plan.

Opperman JJ. 2012. A conceptual model for floodplains in the Sacramento-San Joaquin Delta. *San Francisco Estuary and Watershed Science* 10(3):1-28.

Pasternack GB, Mount JF, Anderson M, Trowbridge W, Jeffres C, Fleenor W. 2004. Observations of a Cosumnes River floodplain. California Water and Environmental Modeling Forum Annual Meeting.

Pestrong R. 1965. The development of drainage patterns on tidal marshes. *Stanford University Publications* 10(2).

Pethick J. 1992. Saltmarsh geomorphology. In Allen JRL, Pye K. Saltmarshes: Morphodynamics, Conservation and Engineering Significance. Pages 41-62. Cambridge University Press, Cambridge.

Ringgold C. 1850. Chart of the Sacramento River from Suisun City to the American River. *Courtesy of the David Rumsey Map Collection.*

Ringgold C. 1850. *A series of charts, with sailing directions.* Washington, DC: JT Towers. *Courtesy of California State Library.*

Roni P, Pess G, Beechie T, Morley S. 2010. Estimating changes in coho salmon and steelhead abundance from watershed restoration: how much restoration is needed to measurably increase smolt production? *North American Journal of Fisheries Management* 30(6):1469-1484.

Rosenfield JA. 2010. *Life History Conceptual Model and Sub-Models for Longfin Smelt, San Francisco Estuary Population.* Aquatic Restoration Consulting.

Rosgen DL. 1994. A classification of natural rivers. *Catena* 22(3):169-199.

Sands A. 1980. *Riparian forests in California: their ecology and conservation: a symposium.* UCANR Publications.

Schiffman PM. 2011. Understanding California grassland ecology. *Fremontia*:12.

Seavy NE, Gardali T, Golet GH, Griggs FT, Howell CA, Kelsey R, Small SL, Viers JH, Weigand JF. 2009. Why climate change makes riparian restoration more important than ever: recommendations for practice and research. *Ecological Restoration* 27(3):330-338.

Seymour G. 1960. *Furbearers of California.* California Department of Fish and Game.

Shuford WD, Gardali T. 2008. *California bird species of special concern: a ranked assessment of species, subspecies, and distinct populations of birds of immediate conservation concern in California. Studies of Western Birds 1.* Western Field Ornithologists, and California Department of Fish and Game.

Simenstad C, Toft J, Higgins H, Cordell J. 2000. *Sacramento/San Joaquin Delta Breached Levee Wetland Study (BREACH).* Wetland Ecosystem Team, Seattle.

Simenstad CA. 1983. *Ecology of estuarine channels of the Pacific Northwest coast: a community profile.* Washington Univ., Seattle (USA). Fisheries Research Inst.

Small SL. 2005. Mortality factors and predators of spotted towhee nests in the Sacramento Valley, California. *Journal of Field Ornithology* 76(3):252-258.

Small SL, Thompson III FR, Geupel GR, Faaborg J. 2007. Spotted Towhee population dynamics in a riparian restoration context. *The Condor* 109(4):721-732.

Smith RD, Ammann A, Bartoldus C, Brinson MM. 1995. *An approach for assessing wetland functions using hydrogeomorphic classification, reference wetlands, and functional indices.* DTIC Document.

Smokorowski K, Pratt T. 2007. Effect of a change in physical structure and cover on fish and fish habitat in freshwater ecosystems-a review and meta-analysis. *Environmental Reviews* 15(NA):15-41.

Sommer T, Baxter R, Herbold B. 1997. Resilience of splittail in the Sacramento–San Joaquin estuary. *Transactions of the American Fisheries Society* 126(6):961-976.

Sommer T, Harrell B, Nobriga M, Brown R, Moyle P, Kimmerer W, Schemel L. 2001a. California's Yolo Bypass: Evidence that flood control can be compatible with fisheries, wetlands, wildlife, and agriculture. *Fisheries* 26(8):6-16.

Sommer T, Nobriga M, Harrell W, Batham W, Kimmerer W. 2001b. Floodplain rearing of juvenile Chinook salmon: evidence of enhanced growth and survival. *Canadian Journal of Fisheries and Aquatic Sciences* 58(2):325-333.

Sommer TR, Harrell WC, Feyrer F. 2013. Large-bodied fish migration and residency in a flood basin of the Sacramento River, California, USA. *Ecology of Freshwater Fish*.

Sommer TR, Harrell WC, Nobriga ML. 2005. Habitat use and stranding risk of juvenile Chinook salmon on a seasonal floodplain. *North American Journal of Fisheries Management* 25(4):1493-1504.

Sommer TR, Harrell WC, Solger AM, Tom B, Kimmerer W. 2004. Effects of flow variation on channel and floodplain biota and habitats of the Sacramento River, California, USA. *Aquatic Conservation: Marine and Freshwater Ecosystems* 14(3):247-261.

Spautz H, Nur N. 2002. *Distribution and abundance in relation to habitat and landscape features and nest site characteristics of California Black Rail (Laterallus jamaicensis coturniculus) in the San Francisco Bay Estuary.* Point Reyes Bird Observatory, Sacramento, CA.

Spautz H, Nur N, Stralberg D, Ralph C, Rich T. 2002. California Black Rail (*Laterallus jamaicensis coturniculus*) distribution and abundance in relation to habitat and landscape features in the San Francisco Bay Estuary. *Bird conservation implementation and integration in the Americas: Proceedings of the Third International Partners in Flight Conference*:20-24.

Spautz H, Nadav N, Stralberg D. 2005. *California Black Rail (Laterallus jamaicensis coturniculus) Distribution and Abundance in Relation to Habitat and Landscape Features in the San Francisco Bay Estuary.* USDA Forest Service Gen. Tech. Rep. PSW-GTR-191.

State and Federal Contractors Water Agency. 2013. *Lower Yolo Restoration Project Draft Environmental Impact Report.* State and Federal Contractors Water Agency, Sacramento, CA.

Takekawa JY, Woo I, Gardiner R, Casazza M, Ackerman JT, Nur N, Liu L, Spautz H. 2011. Avian communities in tidal salt marshes of San Francisco Bay: A Review of functional groups by foraging guild and habitat association. *San Francisco Estuary and Watershed Science* 9(3).

Taylor JR, Cardamone MA, Mitsch WJ. 1990. Bottomland hardwood forests: their functions and values. *Ecological processes and cumulative impacts illustrated by bottomland hardwood wetland ecosystems. Lewis Publ., Chelsea, MI*:13-88.

Teal JM, Aumen NG, Cloern JE, Rodriguez K, Wiens JA. 2009. *Ecosystem Restoration Workshop Panel Report.* CALFED Science Program.

Thompson J. 1957. The settlement geography of the Sacramento-San Joaquin Delta, California. Stanford University, Palo Alto.

Toft JD, Simenstad CA, Cordell JR, Grimaldo LF. 2003. The effects of introduced water hyacinth on habitat structure, invertebrate assemblages, and fish diets. *Estuaries* 26(3):746-758.

Trapp GR. 2011. Vertebrates of California grasslands. *Fremontia*:31.

Turner JL. 1966. *Distribution and food habits of centrarchid fishes in the Sacramento-San Joaquin Delta*. California Department of Fish and Game.

Unwin M. 1997. Fry-to-adult survival of natural and hatchery-produced chinook salmon (*Oncorhynchus tshawytscha*) from a common origin. *Canadian Journal of Fisheries and Aquatic Sciences* 54(6):1246-1254.

U.S. Fish and Wildlife Service. 2012. Pacific Lamprey *Entosphenus tridentata*. Oregon Fish and Wildlife Office, editor. http://www.fws.gov/pacific/Fisheries/sphabcon/Lamprey/pdf/PacificLampreyFactSheet_LongVersion%206-19-12.pdf. Oregon Fish and Wildlife Office, Portland.

U.S. Geological Survey. 2013. National Hydrography Dataset, version 931v210. Reston, Va.: U.S. Geologic Survey. Available at: http://nhd.usgs.gov/index.html. Accessed 2013.

Vasey MC, Parker V T, Callaway JC, Herbert ER, Schile LM. 2012. Tidal wetland vegetation in the San Francisco Bay-Delta estuary. San Francisco Estuary and Watershed Science, 10(2).

Verhoeven JTA, Soons MB, Janssen R, Omtzigt N. 2008. An Operational Landscape Unit approach for identifying key landscape connections in wetland restoration. *Journal of Applied Ecology* 45:1496-1503.

Vogel DA, Marine KR. 1991. *U.S. Bureau of Reclamation Central Valley Project, Draft, Guide to Upper Sacramento River Chinook Salmon Life History*. U.S. Bureau of Reclamation Redding.

Wang R-F, Ateljevich E. 2012. A Continuous Surface Elevation Map for Modeling.

Waples RS, Beechie T, Pess GR. 2009. Evolutionary history, habitat disturbance regimes, and anthropogenic changes: what do these mean for resilience of pacific salmon populations? *Ecology and Society* 14(1):3.

Werner I, Deanovic LA, Connor V, de Vlaming V, Bailey HC, Hinton DE. 2000. Insecticide-caused toxicity to Ceriodaphnia dubia (CLADOCERA) in the Sacramento–San Joaquin River delta, California, USA. *Environmental Toxicology and Chemistry* 19(1):215-227.

West JM, Zedler JB. 2000. Marsh-creek connectivity: fish use of a tidal salt marsh in southern California. *Estuaries* 23(5):699-710.

Weston DP, Lydy MJ. 2010. Urban and agricultural sources of pyrethroid insecticides to the Sacramento-San Joaquin Delta of California. *Environmental science & technology* 44(5):1833-1840.

Wetlands and Water Resources (WWR). 2013. *Natural Communities Mapping of the Cache Slough Complex vicinity compiled from multiple data sources*. Developed for the Cache Slough Complex Conservation Assessment.

Whipple A, Grossinger RM, Rankin D, Stanford B, R A. 2012. *Sacramento-San Joaquin Delta Historical Ecology Investigation: Exploring Pattern and Process*. San Francisco Estuary Institute-Aquatic Science Center, Richmond, CA.

Whisson DA, Quinn JH, Collins KC. 2007. Home range and movements of roof rats (*Rattus rattus*) in an old-growth riparian forest, California. *Journal of Mammalogy* 88(3):589-594.

Whitley SN, Bollens SM. 2014. Fish assemblages across a vegetation gradient in a restoring tidal freshwater wetland: diets and potential for resource competition. *Environmental Biology of Fishes* 97(6):659-674.

Wiens JA. 2002. Riverine landscapes: taking landscape ecology into the water. *Freshwater Biology* 47(4):501-515.

Wiens JA, Hayward GD, Safford HD, Giffen C. 2012. *Historical environmental variation in conservation and natural resource management.* John Wiley & Sons.

Williams JG. 2006. Central Valley Salmon: A Perspective on Chinook and Steelhead in the Central Valley of California. *San Francisco Estuary and Watershed Science* 4(3).

Williams PB, Andrews E, Opperman JJ, Bozkurt S, Moyle PB. 2009. Quantifying activated floodplains on a lowland regulated river: its application to floodplain restoration in the Sacramento Valley. *San Francisco Estuary and Watershed Science* 7(1).

Yoshiyama RM, Fisher FW, Moyle PB. 1998. Historical abundance and decline of chinook salmon in the Central Valley region of California. *North American Journal of Fisheries Management* 18(3):487-521.

Yoshiyama RM, Gerstung ER, Fisher FW, Moyle PB. 2001. Historical and present distribution of Chinook salmon in the Central Valley drainage of California. *Contributions to the Biology of Central Valley Salmonids, Fish Bulletin* 179:71-176.

Young A, Boyle T, Brown T. 1996. The population genetic consequences of habitat fragmentation for plants. *Trends in Ecology & Evolution* 11(10):413-418.